Reservoir Characterization of Tight Gas Sandstones

Reservoir Characterization of Tight Gas Sandstones

Exploration and Development

ALI KADKHODAIE
Earth Sciences Department, Faculty of Natural Sciences,
University of Tabriz, Tabriz, Iran

RAHIM KADKHODAIE
Earth Sciences Department, Faculty of Natural Sciences,
University of Tabriz, Tabriz, Iran

ELSEVIER

Elsevier
Radarweg 29, PO Box 211, 1000 AE Amsterdam, Netherlands
The Boulevard, Langford Lane, Kidlington, Oxford OX5 1GB, United Kingdom
50 Hampshire Street, 5th Floor, Cambridge, MA 02139, United States

ISBN: 978-0-323-90180-2

For Information on all Elsevier publications
visit our website at https://www.elsevier.com/books-and-journals

Publisher: Candice Janco
Acquisitions Editor: Amy Shapiro
Editorial Project Manager: Aera Gariguez
Production Project Manager: Sruthi Satheesh
Cover Designer: Miles Hitchen

Typeset by MPS Limited, Chennai, India

Contents

CHAPTER 1

Tight sandstones: overview, distribution, and global significance

1.1 Introduction

The increasing consumption of energy in the world has drawn a lot of attention to unconventional resources. Unconventional gas resources include shale gas, tight gas, coalbed methane (CBM), and gas hydrates among which tight gas sandstones are of prime importance and have a brilliant future. Permeability is a key parameter to characterize and classify hydrocarbon reservoirs. Tight gas sandstones are low permeability reservoirs that host significant amounts of natural gas. By definition, tight gas sandstones are considered unconventional reservoirs with matrix permeability less than 0.1 mD and matrix porosity less than 10% (Law and Curtis, 2002). The average pore throat size of tight gas sandstones ranges from 0.3 to 2 micrometers (Nelson, 2009). They are called *tight sandstone gas* when we are talking about their gas, and *tight gas sandstone* when we deal with their host rocks. Low permeability, abnormal pressure (overpressure or under-pressure), high gas saturation, the absence of gas-water contacts in down dips, and large areal extension are the important criteria for the accumulation of gas within the tight sandstones of a sedimentary basin. They are mostly distributed in the center or deeper parts of a basin structure with a large regional extension. Therefore, they are sometimes referred to as continuously distributed, basin-centered, or deep basin gas reservoirs. Gas production from the tight sandstones requires a comprehensive knowledge of their geological and physical rock properties such as sedimentary facies, depositional setting, diagenesis, flow units, porosity, permeability, fluids saturation, capillary pressure, and pore size distribution. Due to the low matrix permeability using horizontal wells and stimulation techniques such as hydraulic fracturing has been considered an effective way to produce commercially from tight sandstones. In this chapter, an overview of tight gas sandstone reservoirs, their distribution,

Reservoir Characterization of Tight Gas Sandstones
DOI: https://doi.org/10.1016/B978-0-323-90180-2.00010-4

and global significance are discussed (Zou, 2013; Jiang et al., 2012, 2015b; Moore et al., 2016; Ji et al., 2019).

1.2 Unconventional resources' classification

Generally, unconventional resources are classified into eight groups as follows.

Tight gas: natural gas trapped in low permeability tight sandstones; tight sandstones can produce through hydraulic fracturing or recovery methods like a steam injection.

Shale gas: free gas or absorbed gas trapped in organic-rich shales. Gas shales can produce through hydraulic fracturing techniques.

Coalbed methane: natural gas trapped in coal seams in three forms including dissolved in coalbed water, free gas in coal pores, and gas absorbed on the surface of the coal matrix.

Gas hydrates: A solid form of gas and ice present in its molecular pores. They are abundant on the ocean floors beneath 500 m. Gas hydrates are considered the source of future energy, but currently, they can cause drilling risks and problems.

Heavy oil: High-viscosity crude oil that cannot flow at normal reservoir temperature.

Tar sand (or oil sand): Refers to sandstones containing asphalt, clay, water, and other minerals.

Oil shale: Inflammable shale containing high ash content. To produce from oil shales low-temperature carbonization process is implemented.

Tight oil: Oil accumulated in low permeability carbonates or sandstones. Tight oils can produce through hydraulic or acid fracturing.

Masters (1979) and Gray (1977) proposed the concept of a resources triangle for different types of natural gases found in geological strata (Fig. 1—1). This triangle shows that the gas resources have a log-normal distribution in nature. That is, the conventional gas resources are small in size, but are easy to develop, while unconventional gas resources are found in large volumes in nature, but are difficult to develop. As we move deeper in the triangle, reservoir quality reduces due to decreasing permeability, while the reserves volume increases, dramatically.

Rezaee et al. (2012) based on the in situ permeability values divided the hydrocarbon reservoirs into three categories of conventional reservoirs with a permeability of more than 1 mD, semi-permeable or near-tight

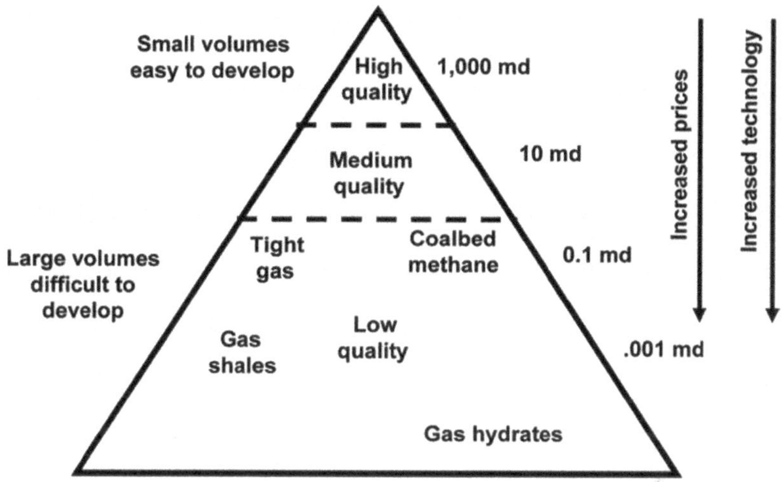

Figure 1–1 Resources triangle for gas reservoirs proposed by Masters (1979) and Gray (1977).

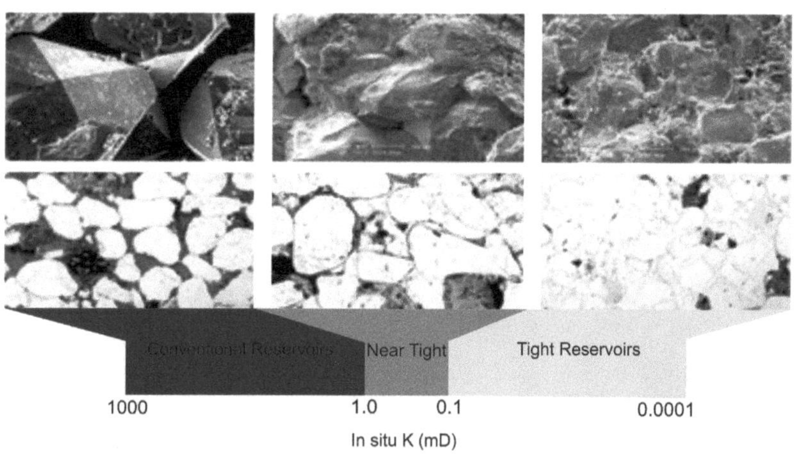

Figure 1–2 Description of conventional, near tight, and tight sandstone reservoirs based on their in situ permeability (Rezaee et al., 2012).

reservoirs with a permeability of 0.1 to 1 mD, and tight reservoirs with a permeability of less than 0.1 mD (Fig. 1–2) (Rezaee et al., 2012).

1.3 Tight gas's worldwide distribution

With increasing oil and gas consumption and depletion of conventional reservoirs, tight gas sandstones are becoming one of the most important resources of the world's energy for the next generation. Shale gas, coalbed methane, and tight gas account for 71.1%, 21.7%, and 7% of global recoverable unconventional gas, respectively (Wang et al., 2016). Tight gas reservoirs form an important part of the natural gas reserves around the globe. Unlike the conventional hydrocarbon reservoirs that produce naturally without the need for stimulation or any special recovery process, tight gas reservoirs, as a type of unconventional reservoirs, due to the low permeability nature are not able to naturally produce gas at a stable and economical production rate, unless they are artificially stimulated by methods such as hydraulic fracturing (Gonfalini, 2005).

Tight gas exploration history back to 1970 in the United States. For the first time, tight gas was produced from the San Juan Basin of Unites States. By 2010, about 23 tight gas basins with more than one hundred thousand wells accounting for gas production of 17.54×10^{10} m^3 were identified by American research groups.

The global resources of tight gas are estimated to be about 7400 tcf (Fig. 1–3). The world's main tight gas resources are distributed in the United States, China, Canada, Russia, Australia, India, Mexico, Africa, and Argentina (Campbell, 2009; McGlade et al., 2012). The tight gas

Figure 1–3 Map showing the global distribution of tight gas resources (Rajput and Kumar Thakur, 2016). Please see the online version to view the color image of the figure.

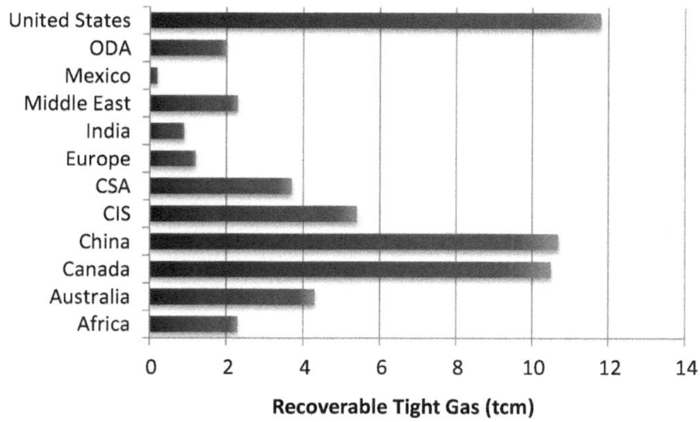

Figure 1–4 Recoverable tight gas resources in the world's top countries. *Note*: *CIS*, Commonwealth of the Independent States; *CSA*, Central and South America; *ODA*, Other Developing Asia.

resources of the other countries and their geologic settings are still understudied. Recoverable tight gas resources in the world's top countries are shown in Fig. 1–4.

1.3.1 The United States

Tight gas accounts for 310 tcf of the technically recoverable gas resources in the United States. According to the Energy Information Administration (EIA), a total of 17% of recoverable gas are hosted in tight gas reservoirs such that the yearly production of tight gas is 6 tcf which is equivalent to 25% of the gas produced in the United States. The main tight gas basins in the United States include Appalachian, Michigan, Northern Great Plains/Williston, Greater Green River, Uinta-Piceance, Wind River, Big Horn, Douglas Creek, Denver, San Juan, Anadarko, Permian, Texas Gulf Coast, Arkoma, East Texas, Black Warrior, Ozona, Sonora, Edwards Lime, Cotton Valley (sweet), Cotton Valley (sour) and Ouachita (Fig. 1–5). The first production from tight sandstones was developed in the San Juan Basin of New Mexico in an area of 9325 km^2 with porosity varying from 5.8% to 7.6% and permeability ranging from 0.01 to 0.15 mD. The Late Cretaceous Mesaverde (Chacra sandstone member), the Early Cretaceous Dakota, and the Late Cretaceous Pictured Cliff formations of the Balnco field are the main reservoir units of the San Juan Basin with a geological reserve of 0.9×10^{12} m^3, respectively. The Red Desert Basin of Wyoming (also known as the Great Divide) with an

Figure 1–5 Distribution map of major tight gas reservoirs in the United States (Chinedu Ozobeme, 2006).

area of 24,100 km^2 (porosity varying from 3% to 7% and permeability falling below 1 mD) is estimated to host 1.4×10^{12} m^3 of natural gas. The Late Cretaceous sandstones of the Almond, Lance, Ericson, Rock Springs, and Blair formations along with the Paleocene Fort Union formation are the major tight gas reservoirs of the Red Desert Basin.

The Greater Green River Basin with a geological gas reserve of 0.26×10^{12} m^3 covers an area of 51,022 km^2 with porosity varying from 8% to 12% and permeability ranging from 0.1 to 0.9 mD. The main reservoirs of the Greater Green River Basin include the Late Cretaceous U. Almond, Blair, and Mesaverde formations of Echo Spring field along with the Frontier formation of the La Barge platform. The major tight gas formations and fields of the United States Basins are listed in Table 1–1. The historical and predicted proportion of various gases production from 1990 to 2035 in the United States is shown in Fig. 1–6. As is seen, it is expected that tight gas with a proportion of 21% in 2035 will receive the second rank of total gas output after shale gas.

1.3.2 China

The tight gas exploration in China was first reported in 1971 through the discovery of the Zhongba field in the western Sichuan Basin. The main

Table 1–1 Major tight gas fields and formations in the United States.

Basin	Field	Formation	Geological age
Arkoma	Kinta	Spiro sandstone	L. Atokan
Anadarko	Ellis Ranch	Cleveland	U. Pennsylvanian
	Strong City	Cherokee	Pennsylvanian
	Hemphill	Granite wash	Pennsylvanian
Appalachian	Clinton	Clinton medina	L. Silurian
	Berea	Berea sandstone	U. Devonian
Denver	Wattenberg	Muddy (J) sandstone	L. Cretaceous
Fort Worth	Boonesville	Davis sandstone	Pennsylvanian
Gulf Coast	J.C. Martin	Wilcox lobo	Paleocene
	J.C. Martin	Wilcox deltaic	Eocene
	McAllen Ranch	Vicksburg	Oligocene
Great Plains	Bowdoin Dome	Bowdoin sandstone	Cretaceous
Green River	La Barge Platform	Frontier	U. Cretaceous
	Echo Springs	U. Almond	U. Cretaceous
	Echo Springs	Blair	U. Cretaceous
	Echo Springs	Mesaverde	U. Cretaceous
Hugoton Embayment	Panhandle	Chase group	Permian
Maverick	Dos Hermanos	Olmos	U. Cretaceous
North Louisiana	Carthage	Cotton valley	U. Jurassic
	Carthage	Travis peak	L. Cretaceous
Permian	Pecos Slope	Abo	L. Permian
	Indian Basin	Upper Penn. Dolostone	U. Pennsylvanian
	Pitchfork Ranch	Morrow	L. Pennsylvanian
	Spraberry	Spraberry (Oil)	Permian
Piceance	Douglas Creek	Dakota sandstone	L. Cretaceous
	Shire Gulch	Mancos B Shale	U. Cretaceous
	Plateau	Mesaverde group	U. Cretaceous
San Juan	Blanco	Dakota	L. Cretaceous
	Blanco	Mesaverde group Ss.	U. Cretaceous
	Blanco	Pictured cliff Ss.	U. Cretaceous
	Blanco	Charca Ss.	U. Cretaceous
Uinta	Altamount	Green river	Eocene
	Douglas Creek	Douglas creek arch	Eocene
Val Verde	Ozona Sonora	Canyon	U. Pennsylvanian
Wind River	Riverton Dome	Frontier Ss.	U. Cretaceous

tight gas-bearing basins in China include Ordos, Sichuan, Tarim, North Songliao, South Songliao, Juggar, Jianghan, Jiuquan, East Sea, Turpan-Hami (Tuha), Qaidam, Chuxiong, Bohai Bay, Taixi, Yingqiog, and Pearl River Mouth Basin (Fig. 1–7). The upper Paleozoic sandstones of Ordos and the Triassic Xujiahe sandstones of Sichuan along with the Jurassic-Lower Cretaceous sandstones of Tarim Basins host about 40% of proven natural gas in China (equivalent to 3.6×10^{12} m^3) (Hu et al., 2010; Dai et al., 2012; Wang et al., 2014).

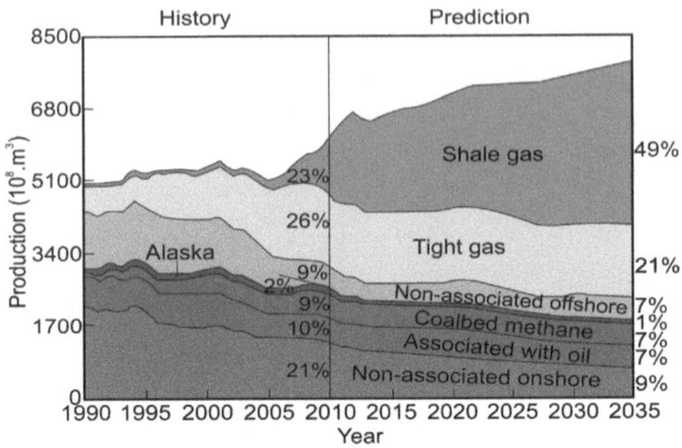

Figure 1–6 Chart showing the historical (1990–2010) and predicted (2010–35) proportion of various gases produced in the United States (Dai et al., 2012).

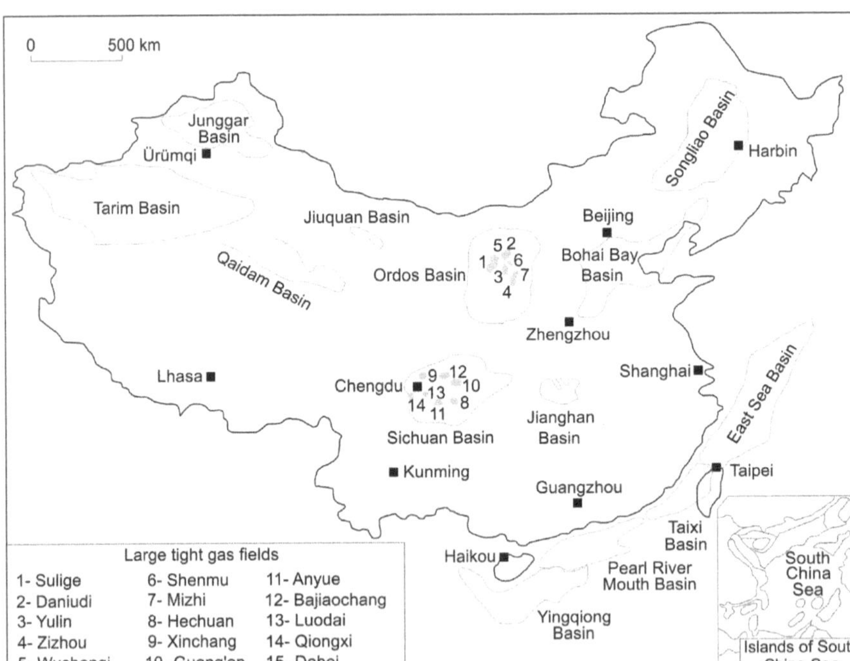

Figure 1–7 Distribution map of major tight gas reservoirs in China (Dai et al., 2012).

Tight gas sandstones in the Sulige gas field (Ordos Basin), the western gas fields of the Sichuan Basin, and the Kuqa depression in the Tarim Basin are associated with the coal beds. The coal measures are rich in organic matter, widely distributed, and have high gas production intensity. Accordingly, coal measures and lacustrine mudstone have been considered the main source rocks of tight gas in the central and western gas fields of China. The proximity of coal measures to reservoir rocks, short migration distances, and widely distributed sandstones has increased the expulsion efficiency of source rocks and gas accumulation in the prolific basins (Yang et al., 2008; Jiang et al., 2015a).

1.3.3 Canada

Unconventional gas plays in Canada have widely been considered exploration and development targets since 2005. Western Canada and the Rocky Mountains' gas-prone sedimentary basins contain significant amounts of unconventional gas (Law, 2002). There are nearly 20 tight gas basins across the Rocky Mountains Basin Group and Gulf coast (Lei et al., 2010). The giant Elmworth deep tight gas of Alberta Basin is the largest field in North America covering an area of 13,000 km^2 with a geological reserve of 4.8×10^{12} m^3. It produces from the Lower Cretaceous tight sandstones of Cadotte, Falher, Bluesky, and Cadomin formations with average porosity of 8% and permeability less than 1 mD. The Western Canada Deep Basin hosts the deepest basin-centered tight gas reserves of the Triassic Montney siltstones. Montney reservoir of up to 300 m thick with a geological reserve of 449 tcf gas covers an area of 100 km wide and hundreds of kilometers long (Jiang et al., 2015a). Fig. 1–8 represents the distribution map of major tight gas reservoirs in Canada.

1.3.4 Australia

Australia is one of the pioneer countries in the exploration and production of tight gas sandstones. The location map of the major unconventional resources of Australia including tight gas reservoirs is shown in Fig. 1–9. The Warro tight gas field of Perth Basin is one of the largest unconventional gas resources in Western Australia (WA). It can supply more than 10% of WA gas consumption. The mail-producing tight gas fields occur in Perth Basin and Cooper Basin. The major tight gas reservoirs of Australian basins are outlined as follows (Orsini et al., 2011; Johnson et al., 2019; Kadkhodaie et al., 2021).

1.3.4.1 Queensland

Laura Basin: Dalrymple Sandstone

Maryborough Basin: Maryborough Formation

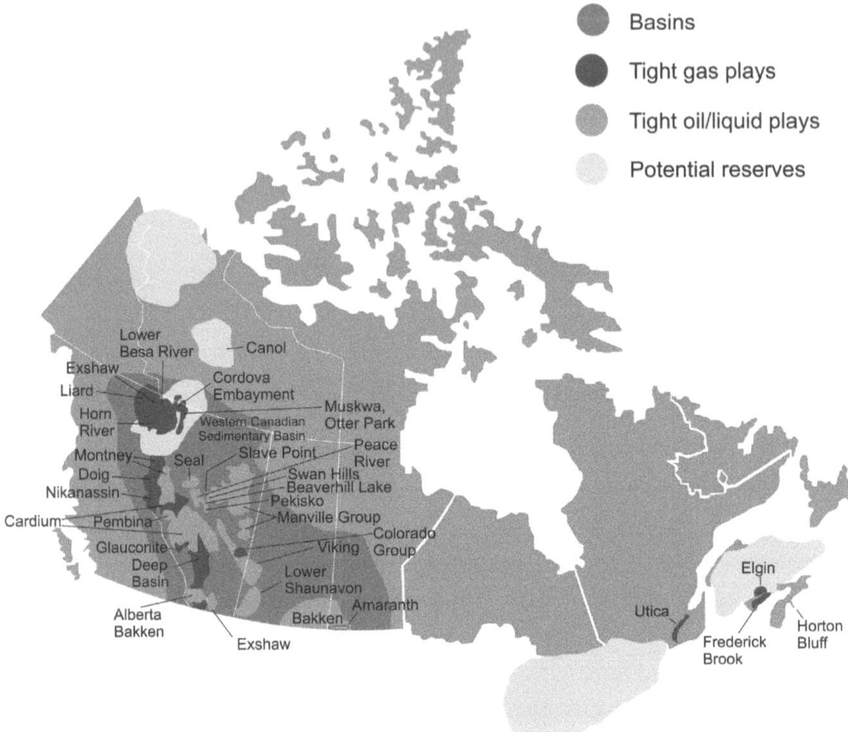

Figure 1–8 Distribution map of major tight gas reservoirs in Canada. Please see the online version to view the color image of the figure. *Map source: From ConocoPhillips Canada (2013); Hamblin (2010); Ontario Geological Survey (2009); PacWest Consulting Partners (2013); US Energy Administration (2011); Yukon Government (2015).*

Bowen Basin: Tinowon Formation and Bandanna Formation

Cooper Basin: Toolachee Formation, Epsilon Formation, and Patchawarra Formation

Galilee Basin: Lake Galilee Sandstone

Adavale Basin: Log Creek Formation and Lissoy Sandstone

Georgina Basin: Arrinthrunga Formation.

1.3.4.2 New South Wales

Sydney Basin: Narrabeen Group, Bulgo Sandstone, Colo Vale Sandstone, Illawarra Coal Measures, and Shoalhaven Group.

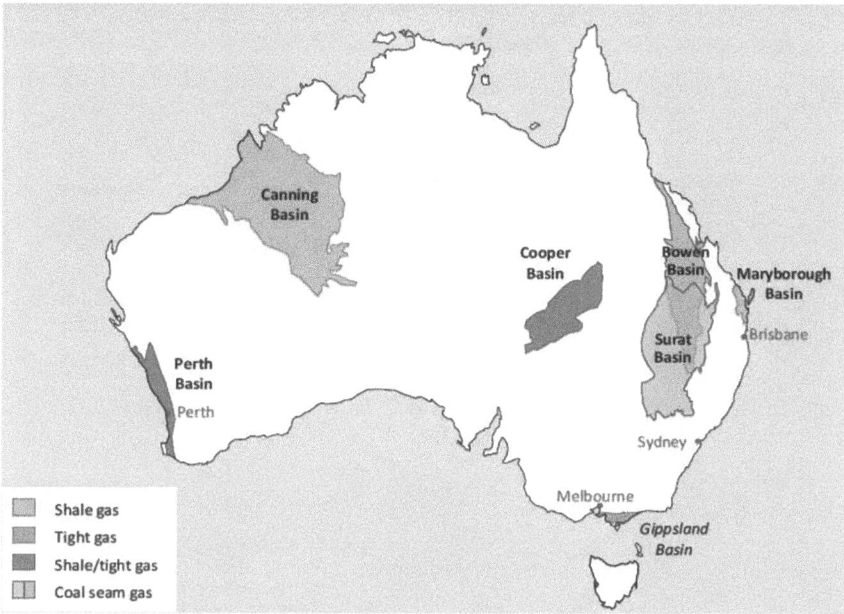

Figure 1–9 Distribution map of the major unconventional resources of Australia including tight gas reservoirs. Please see the online version to view the color image of the figure. *Map from International Energy Agency (IEA).*

1.3.4.3 South Australia

Cooper Basin: Toolachee Formation, Epsilon Formation, and Patchawarra Formation

Warburton Basin: Pando Formation, Dullingari Group, Kalladeina Formation, and Mooracoochie Volvcanics

Officer Basin: Observatory Hill Formation, Ouldburra Formation, Narana Formation, and Dee Dee Mudstone.

Otway Basin: Pretty Hill Formation.

1.3.4.4 Western Australia

Northern Perth Basin: Yarragadee Formation, Dongara/Wagina Sandstone, Irwin River Coal Measure, High Cliff Sandstone

Canning Basin: Laurel Formation.

Bonaparte Basin: Milligans Formation and Bonaparte Formation.

1.3.4.5 Northern Territory

Onshore Bonaparte Basin: Milligans Formation and Bonaparte Formation

Georgina Basin: Arthur Creek Formation

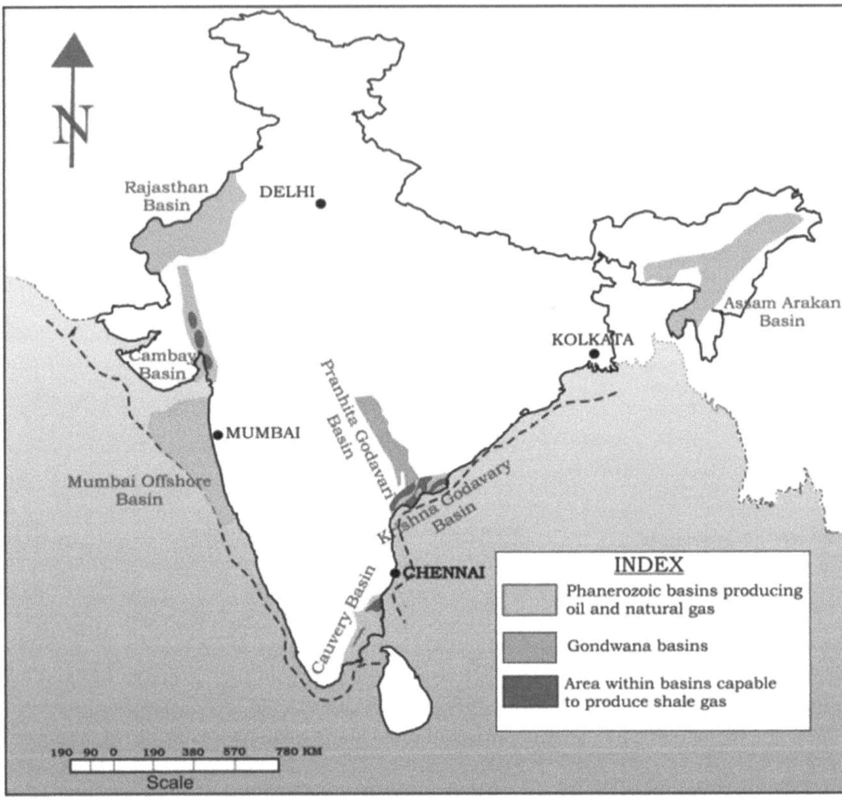

Figure 1–10 Distribution map of major tight gas basins in India (Pratim Chakraborty et al., 2013). Please see the online version to view the color image of the figure.

McArthur Basin/Beetaloo Sub-basin: Velkerri Formation, Coxco Dolostone, Bessie Creek Sandstone, and Moroak Sandstone.

Amadeus Basin: Pacoota Sandstone and Stairway Sandstone.

Ngala Basin: Mount Eclipse Sandstone.

1.3.5 India

Tight gas sandstones of India occur in most of the producing basins including Krishna-Godavari (KG), Cauvery, Cambay, Bengal, and Vindhyans Basins (Fig. 1–10).

In the KG basin, the Cretaceous Raghavapuram and the deeper Golapalli Formations are the gas-bearing tight sandstones. Mechanical compaction coupled with authigenic clay minerals has significantly reduced the reservoir porosity and permeability. The Cauvery Basin contains the Early Cretaceous Andimadam and the Late Cretaceous Bhuvanagiri tight sandstone Formations. The wedged out area of the Mukta and Bassein Formations in the Bombay high (the southern plunge) is composed of tight reservoirs. In the Cambay Basin, the thick sequences of the Eocene deposits host considerable resources of tight gas. In some localities of the Assam and Arakan Basin, the Tertiary sequences of the lower and middle Tipams Formation together with the Barail and Kopilies Formations in the Geleki field are tight sandstones in nature and require hydraulic fracturing for commercial production (Kumar and Shandilya, 2013).

1.3.6 Middle East and North Africa

Due to the large production from the giant and conventional oil and gas fields, unconventional resources have not been seriously considered for development and production in the Middle East. Oman, Saudi Arabia, Egypt, Algeria, Iraq, Jordan, Syria, Afghanistan, and Iran are the main countries that host considerable resources of tight gas sandstones in the Middle East and North Africa (Table 1–2). In the Caspian Sea Basin, the Middle Pliocene series of the Pereriv formation (also known as the Fasila) are deep water-tight sandstones hosting considerable amounts of natural gas. Apart from Iran and Egypt, major tight gas sandstone resources in the Middle East are hosted in the Paleozoic sequences.

1.3.7 Central Europe

The main source of tight gas in Europe is the Permian Rotliegende sandstones (Fig. 1–11), whose development began in 1959 with the discovery of the Groningen structure onshore Holland. Then, in 1965, with the discovery of West Soul, it was proven that similar Rotliegende sediments had spread across the southern North Sea. About 300 gas fields and 75 Tcf of recoverable natural gas have been discovered in the southern North Sea, and about 260 gas fields and 150 Tcf of natural gas have been found onshore in the Netherlands, Germany, and Poland (Hartley, 2012). The Central European Basin system includes several sub-basins with a large number of Plaeozoic gas fields that extends from the North Sea through the Netherland and Germany to Poland (Gaupp et al., 2008; Lokhorst et al., 1998). Salt tectonics has strongly affected the basin. A summary of tectonic and sedimentary evolution and fluid dynamics of the Central

Table 1–2 Major tight sandstone formations in the Middle East and North Africa (Shehata et al., 2010).

Country	Field name	Geologic age	Reservoir properties		Applied technology
			φ (%)	*K* (mD)	
Oman	Huge Khazzan and Makarem gas field	Ordovician	7	0.1	• Wide azimuth seismic • Directional drilling • Underbalanced drilling • Hydraulic-fracture
Saudi Arabia	Gawaher	Ordovician/ Silurian	12	>1	• 3D seismic • Horizontal drilling • Multi hydraulic frac stages
Saudi Arabia	Mushayab	Ordovician	4–6	0.001–0.008	• Horizontal drilling • Fracturing Horizontal wells • Multi hydraulic frac stages
Egypt	Obaiyed	Middle Jurassic	7–13	0.1–600	• Hydraulic-fracture • Directional drilling
Egypt	Abu Gharadig	Upper cretaceous	8.5	0.01–200	• Hydraulic-fracture • Directional drilling
Algeria	Teguentour	Silurian Lower Devonian	20.1	<1	• Underbalanced drilling • Horizontal drilling • Fracturing Horizontal wells • Multi hydraulic frac stages
Algeria	Krechba	Silurian Lower Devonian	8.5	<1	• 2D Seismic • Horizontal drilling • Geosteering • Multi hydraulic frac stages

Algeria	In Salah	Silurian Lower Devonian	10	< 1	• 3D multi-azimuth seismic survey • Nuclear magnetic resonance (NMR) • Directional drilling • Hydraulic-fracture • Improved proppants
Iraq	Akkas	Upper Ordovician	7.6	0.13	• Wide azimuth seismic • Underbalanced drilling • Directional drilling • Multi hydraulic frac stages
Jordan	Risha	Upper Ordovician	3–7	< 0.124	• Directional drilling • Hydraulic-fracture
Syria	Arak	Carboniferous	13	< 1	• 2D Seismic • Directional drilling • Coiled tubing drilling • Acidizing the formation matrix
Iran	Moghan	Oligo-Miocene Zivar & Ojagh Gheshlagh	1–12	< 1	• Hydraulic-fracture shall be employed

Source: From Shehata, A., Aly, A., Ramsey, L., 2010. Overview of tight gas field development in the Middle East and North Africa region. Society of Petroleum Engineers, SPE 126181.

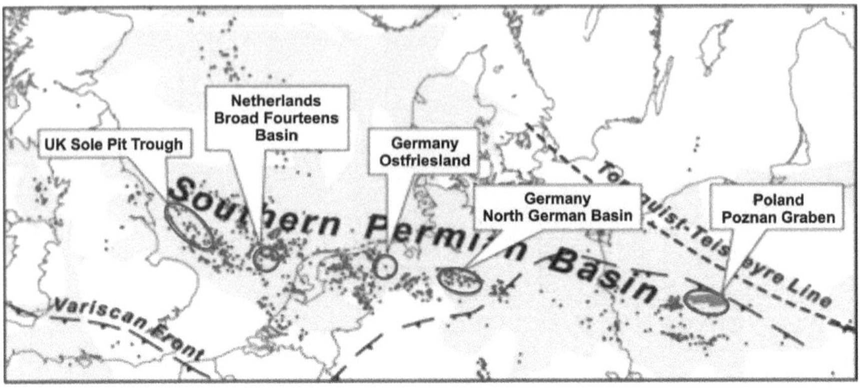

Figure 1–11 The Rotliegendes Basin and active tight gas areas. *From Dr. Andy Sims, Merlin Energy Resources.*

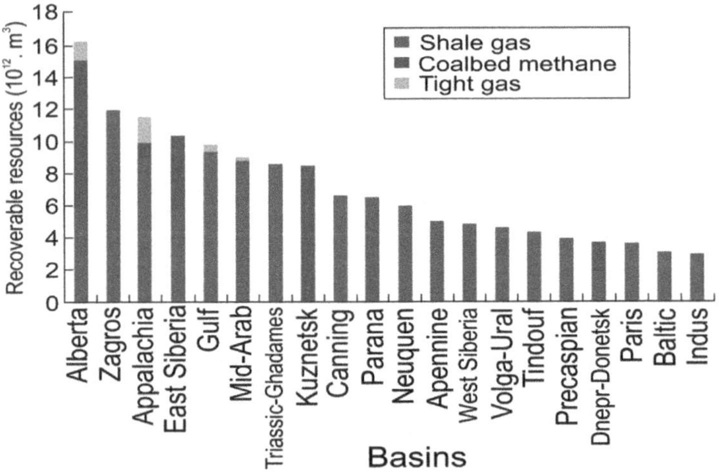

Figure 1–12 Recoverable unconventional gas in the top 20 sedimentary basins (Wang et al., 2016). Please see the online version to view the color image of the figure.

European Basin System has been provided by Littke et al. (2008). The giant Groningen gas field in the northeast of the Netherlands is the largest of these structures. Tight gas sandstone reservoirs are especially present in the Upper Permian Rotliegend Formation of the Netherlands and northern Germany as well as Upper Carboniferous (Pennsylvanian). The low permeability nature of these sandstones is attributed to the deep burial and their complex cementation history. The Upper Carboniferous coal-bearing strata are the main source of gas accumulations (Littke et al., 2011).

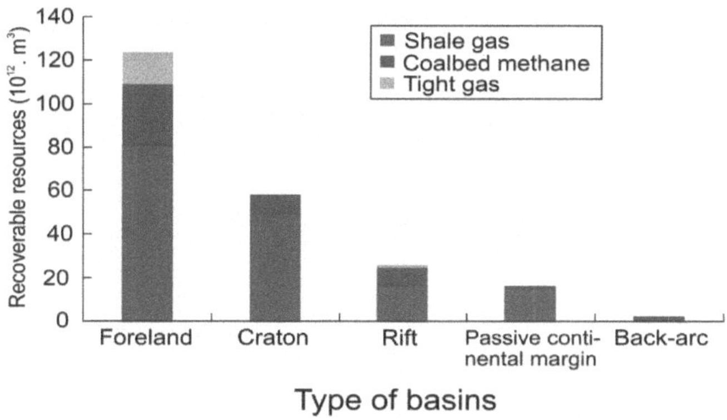

Figure 1—13 Recoverable resources in the tectonic basins hosting unconventional gas (Wang et al., 2016). Please see the online version to view the color image of the figure.

Although sandstone reservoirs are the main hosts of tight gas, other sedimentary rocks such as siltstone and limestones can be considered as a tight gas reservoirs in the case of having permeability less than 0.1 mD, irrespective of their sedimentary environment. The main sedimentary environments of tight sands include delta fans, alluvial fans, fluvial channels, shelf margins, and submarine fans (Law, 2002; Holditch, 2006; Rushing et al., 2008; Kennedy et al., 2012; Wei and Xu, 2015).

Recoverable unconventional gas resources in the world's top 20 sedimentary Basins are shown in Fig. 1—12. As is seen, Appalachia, Alberta, Gulf, and Mid-Arab Basins contain the largest amount of the world's tight gas resources. From a tectonics standpoint, the main sedimentary basins hosting tight sands are foreland basins (e.g., Appalachian and Alberta foreland Basins) (Fig. 1—13). From the geologic time point of view, the main tight gas reservoirs are formed in the Cretaceous, Silurian, Carboniferous, and Permian series of strata (Fig. 1—14).

1.4 Tight gas accumulation models

Most of the tight gas sandstone resources occur in basin-centered or deep basin settings (Fig. 1—15). By definition, the basin-centered tight gas accumulations are abnormally pressured and have low permeability and deep basin (> 1500 ft) gas reservoirs lacking a downdip gas-water contact (Masters, 1979). The alternative names used to describe basin-centered gas reservoirs are "flip-type" syncline gas reservoir, tight gas sandstones, a

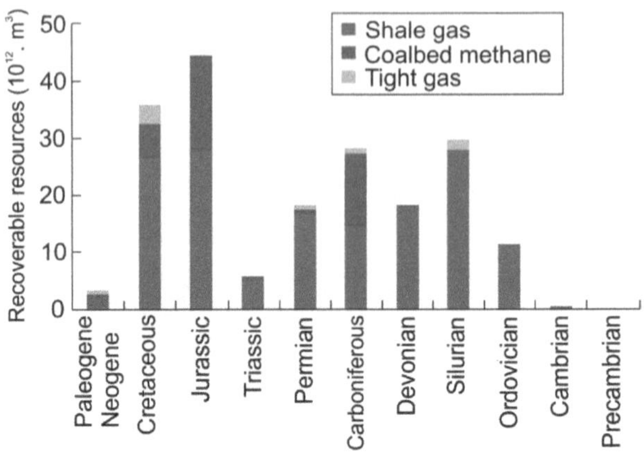

Figure 1–14 Histograms showing the recoverable unconventional gas in geologic time scale (Wang et al., 2016). Please see the online version to view the color image of the figure.

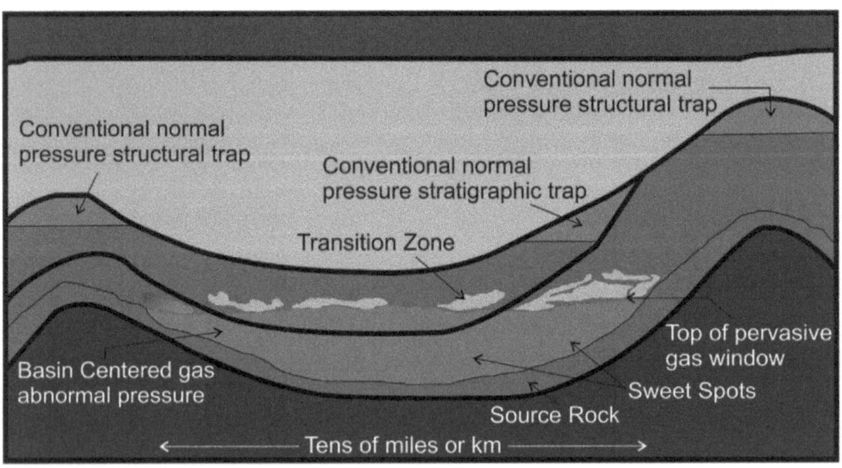

Figure 1–15 Schematic diagram of BCGAs (Schenk and Pollastro, 2002). *BCGAs*, Basin-centered gas accumulations.

continuous gas reservoir, source-contacting gas, and pervasive sandstone reservoirs (Walls, 1982; Rose et al. 1984; Law and Dickinson 1985; Zhang, 2006; Zou, 2013). Apart from the basin-centered settings, they can also occur in the fluvial systems, alluvial fans, delta fans, submarine fans, barrier bars, and tidal bars (Holditch, 2006). Tight gas sandstones can form in different tectonic settings ranging from compressional to extensional basins such as foreland, craton, rift, passive continental margin, and back-arc basins. Tight sands can also form as a result of deep burial diagenesis (e.g., tight sandstones of the Moghan area, Northwestern Iran, and tight gas sandstones of the Shurijeh Formation, Northeastern Iran).

The source rocks of tight gas in basin-centered accumulations are thought to be adjacent to their reservoir rocks. This way, gas expulsion from a mature source rock drives the gas into the low permeability sandstone reservoir (Meckel and Thomasson, 2008). Source rocks should be rich in total organic carbon, thick enough, and lie in the gas generation window. The best tight gas targets for exploration and production are those adjacent to the organically rich source rocks. Both structural and stratigraphic traps can explain the primary controls on accumulations of tight gas resources. Different types of tight gas sandstone traps along with the trapping/migration agents are illustrated in Fig. 1–16.

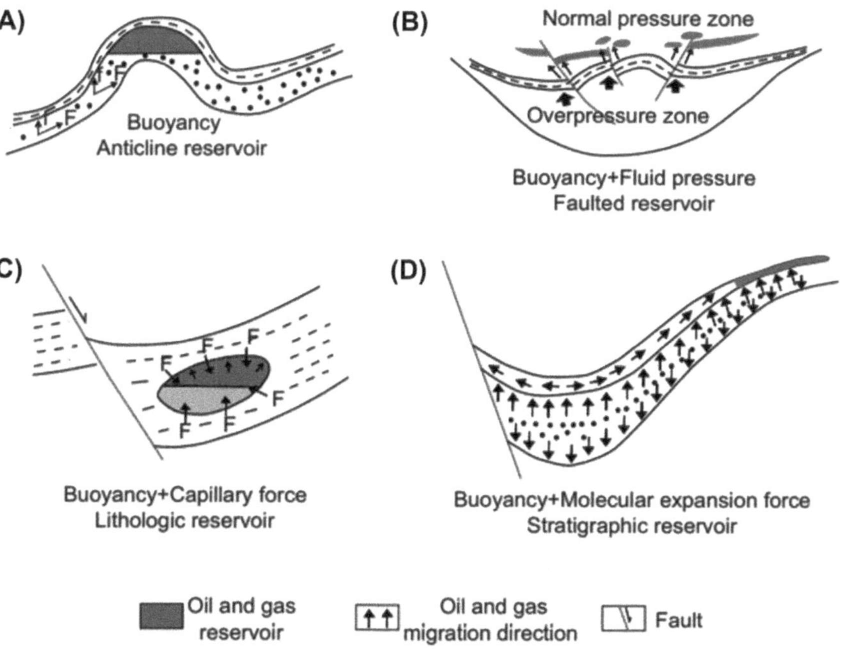

Figure 1–16 Different types of tight gas sandstone traps: (A) anticlinal, (B) fault-related, (C) stratigraphic lens, and (D) stratigraphic pinch-out trap (Jiang et al., 2015b).

References

Campbell, I, 2009. An overview of tight gas resources in Australia, Petroleum Exploration Society of Australia NEWS, 95–100.

Chinedu Ozobeme, C., 2006. Evaluation of water production in tight gas sands in the Cotton Valley Formation in the Caspiana, Elm Grove and Frierson Fields. Master of Science thesis, Texas A & M University, 244p.

Dai, J., Ni, Y., Wu, X., 2012. Tight gas in China and its significance in exploration and exploitation. Petroleum Exploration and Development 39 (3), 277–284.

Gaupp, R., Möller, P., Lüders, V., di Primio, R., Littke, R., 2008. Fluids in sedimentary basins: an overview. In: Littke, R., Bayer, U., Gajewski, D., Nelskamp, S. (Eds.), Dynamics of Complex Intracontinental Basins. The Central European Basin System. Springer, Berlin, pp. 345–359.

Gonfalini, M., 2005. Formation evaluation challenges in unconventional tight hydrocarbon reservoirs SPE Italian Section.

Gray, J.K. 1977. Future gas reserve potential Western Canadian Sedimentary Basin. Third National Technology Conference Canadian Gas Association.

Hartley, R., 2012. What is tight gas? Since its production is tied to technological developments, well cost, and fraccing gas prices, tight gas is an economic issue. Geo Science Explained 9 (3).

Holditch, S.A., 2006. Tight gas sands. Journal of Petroleum Technology 86–93 (June).

Hu, G., Li, J., Shan, X., Han, Z., 2010. The origin of natural gas and the hydrocarbon charging history of the Yulin gas field in the Ordos Basin, China. International Journal of Coal Geology 81, 381391.

Ji, G., Jia, A., Meng, D., Guo, Z., Wang, G., Cheng, L., et al., 2019. Technical strategies for effective development and gas recovery enhancement of a large tight gas field: a case study of Sulige gas field, Ordos Basin, NW China. Petroleum Exploration and Development 46 (3), 629–641.

Jiang, Fu-Jie Kennedy, R.L., Knecht, W.N., Georgi, D.T., 2012. Comparison and contrasts of shale gas and tight gas developments, North American experience and trends. In: SPE Paper 160855, Presented at the SPE Saudi Arabia Section Technical and Exhibition, Al-Khobar, Saudi Arabia, 8–11 April.

Jiang, Z.X., Li, Z., Li, F., Pang, X.Q., Yang, W., Liu, L.F., et al., 2015a. Tight sandstone gas accumulation mechanism and development models. Petroleum Science 12 (4), 587–605.

Jiang, X., Zhi-qiang, M., Yan, J., 2015b. Tight gas sandstone reservoirs evaluation from nuclear magnetic resonance (NMR) logs: case studies. Arabian Journal for Science and Engineering 40 (40), 1223–1237.

Johnson, R.L., Zhong, R., Nguyen, L., 2019. Well trajectory, completion and fracture design changes improve execution for deep unconventional tight gas targets in the Cooper Basin, Australia." Paper presented at the SPE Hydraulic Fracturing Technology Conference and Exhibition, The Woodlands, Texas, USA, February 2019. https://doi.org/10.2118/194366-MS.

Kadkhodaie, R., Kadkhodaie, A., Rezaee, R., 2021. Study of pore system properties of tight gas sandstones based on analysis of the seismically derived velocity deviation log: a case study from the Perth Basin of western Australia. Journal of Petroleum Science and Engineering 196, 108077.

Kennedy, R.L., Knecht, W.N., Georgi, D.T., 2012. Comparison and contrasts of shale gas and tight gas developments, North American experience and trends. In: SPE Paper 160855, Presented at the SPE Saudi Arabia Section Technical and Exhibition, Al-khobar, Saudi Arabia, 8–11 April.

Kumar, T., Shandilya, A., 2013. Tight reservoirs: an overview in Indian context. In 10th Biennial International Conference and Exposition, Kochi, 8p.
Law, B.E., 2002. Basin-centered gas systems. AAPG Bulletin 86 (11), 1891–1919.
Law, B.E., Curtis, J.B., 2002. Introduction to unconventional petroleum systems. AAPG Bulletin 86, 1851–1852.
Law, B.E., Dickinson, W.W., 1985. Conceptual model for origin of abnormally pressured gas accumulations in low-permeability reservoirs. AAPG Bulletin 69 (8), 1295–1304.
Lei, Q., Wan, Y., Li, X., Hu, Y., 2010. A study on the development of tight gas reservoirs in the USA. Natural Gas Industry 30 (1), 45–48.
Littke, R., Bayer, U., Gajewski, D., Nelskamp, S., 2008. Dynamics of complex intracontinental basins. The example of the Central European Basin System. Springer-Verlag, Berlin-Heidelberg.
Littke, R., Krooss, B., Uffmann, A.K., Schulz, H.M., Horsfield, B., 2011. Unconventional gas resources in the Paleozoic of Central Europe. Oil & Gas Science and Technology – Rev. IFP Energies nouvelles 66 (6), 953–977.
Lokhorst A., Adlam K., Brugge J.V.M., David P., Diapari L., Fermont W.J.J., et al., 1998. NW European Gas Atlas – Composition and Isotope Ratios of Natural Gases, CD-ROM, Nederlands Inst. Voor Toegepaste Geowetenschappen TNO, Haarlem.
Masters, J.A., 1979. Deep basin gas trap, western Canada. AAPG Bulletin 63, 152–181.
McGlade, C., Speirs, J., Sorrell, S., 2012. Unconventional gas- a review of estimates. Imperial College Centre for Energy Policy and Technology (ICEPT) Working Paper, pp. 1–29.
Meckel, L.D., Thomasson, M.R., 2008. Pervasive tight gas sandstone reservoirs: an overview. In: Cumella, S.P., Shanley, K.W., Camp, W.K. (Eds.), Understanding, Exploring, and Developing Tight Gas Sands, AAPG Hedberg Series, vol. 3, pp. 13–27. Tulsa, OK.
Moore, W.R., Zee Ma, Y., Pirie, I., Zhang, Y., 2016. Tight gas sandstone reservoirs, part 2: petrophysical analysis and reservoir modeling, chapter 15 in book. In: Zee Ma, Y., Holditch, S.A. (Eds.), Unconventional Oil and Gas Resources Handbook, Evaluation and Development. Gulf Professional Publishing, Oxford, United States, pp. 429–448.
Nelson, P.H., 2009. Pore-throat sizes in sandstones, tight sandstones, and shales. American Association of Petroleum Geologists Bulletin 93 (3), 329–340.
Orsini, C., Rezaee, Wilson, M., 2011. Factors controlling tight gas sandstone reservoir quality in the whicher range gas field, southern Perth Basin, western Australia. Journal of the Australian Petroleum Production & Exploration Association (APPEA) 51 (2), 741.
Pratim Chakraborty, P., Sharma, R., Basu Roy, S., 2013. A key role played by hydrocarbon industry in Indian economy and the road ahead. International Journal of Advancement in Earth and Environmental Sciences 1 (1), 54–62.
Rajput, S., Kumar Thakur, N., 2016. Geological Controls for Gas Hydrate Formations and Unconventionals. Elsevier Press, Amsterdam, Netherlands, p. 384.
Rezaee, R., Saeedi, A., Clennell, B., 2012. Tight gas sands permeability estimation from mercury injection capillary pressure and nuclear magnetic resonance data. Journal of Petroleum Science and Engineering 88–89, 92–99.
Rose, P.R., Everett, J.R., Merin, I.S., 1984. Possible basin centered gas accumulation, Raton Basin, southern Colorado. Oil and Gas Journal 82 (40), 190–197.
Rushing, J.A., Newsham, K.E., Blasingame, T.A., 2008. Rock typing - keys to understanding productivity in tight gas sands. SPE 114164.
Schenk, C.J., Pollastro, 2002. Natural gas production in the United States. US Geological Survey Fact Sheet FS-113-01, 2p.
Shehata, A., Aly, A., Ramsey, L., 2010. Overview of tight gas field development in the Middle East and North Africa Region. Society of Petroleum Engineers, SPE 126181.

Walls, J.D., 1982. Tight gas sands-permeability, pore structure, and clay. Journal of Petroleum Technology 34 (11), 2707–2714.
Wang, H., Ma, F., Tong, X., Liu, Z., Zhang, X., Wu, Z., et al., 2016. Assessment of global unconventional oil and gas resources. Petroleum Exploration and Development 43 (6), 925–940.
Wang, W., Wang, W., Yai, J., Cai, Y., 2014. Tight gas reservoir accumulation conditions and distribution patterns during the upper palaeozoic of the east-central ordos basin, central China. Energy Exploration & Exploitation 32 (4), 657–672.
Wei, Y., Xu, J., 2015. Development of liquid-rich tight gas sand plays – granite wash example. In: Ma, Y.Z., Holditch, S., Royer, J.J. (Eds.), Unconventional Resource Handbook: Evaluation and Development. Elsevier.
Yang, H., Fu, J., Wei, X., Liu, X., 2008. Sulige field in the ordos basin: geological setting, field discovery and tight gas reservoirs. Marine and Petroleum Geology 25, 387–400.
Zhang, J.C., 2006. Source-contacting gas: derived from deep basin gas or basin-centered gas. Natural Gas Industry 26 (2), 46–48.
Zou, C., 2013. Unconventional Petroleum Geology. Elsevier, p. 373.

CHAPTER 2

Sedimentary features of tight gas reservoirs

2.1 Introduction

As with conventional sandstones, tight sandstone reservoirs are characterized by both depositional and post-depositional (diagenetic) characteristics. Regardless of reservoir properties, tight sandstones are comparable to conventional sandstones in terms of different sedimentary parameters, including composition, depositional texture, and structure. In fact, tight sandstones inherit most of the primary depositional environment attitudes, which are recorded and reflected as compositional, textural, and structural parameters in their sedimentary facies. Therefore, before any judgment about the reservoir properties of tight sandstones, it is essential to reach an understanding of their sedimentary facies and environment. This is important at least in two aspects, (1) sedimentary characteristics have the first control on reservoir properties, reservoir heterogeneity, and distribution of reservoir zones; (2) diagenetic history of tight sandstone reservoirs (i.e., severity of diagenetic processes and distribution of diagenetic products) is mainly controlled by primary depositional texture and composition of these reservoirs.

2.2 Sandstones: from sedimentary characteristics to reservoir properties

Sandstone is a clastic sedimentary rock that in terms of sedimentological characteristics consists of medium-sized sedimentary grains (2–0.0625 mm), in the Udden and Wentworth (1922) scale (Table 2–1). Hence, sand grains are finer than gravel (>2 mm) in rutaceous rocks (conglomerates and breccias) and coarser than silt and clay (<0.0625 mm) in fine-grained and argillaceous sedimentary rocks (shale, siltstone, and claystone). Sedimentary grains in sandstones are formed as a result of weathering and erosion of the rocks in the Earth's crust, transporting them mostly by water and wind, and eventually deposition in a variety of sedimentary environments (continental to marine). On the other hand, physical, chemical, and biological conditions of

DOI: https://doi.org/10.1016/B978-0-323-90180-2.00005-0

Table 2–1 Grain size scale of Udden-Wentworth for classification of clastic sediments.

Grain diameter		Class size	Sediment	Rock
Phi	mm			
−8	> 256	Boulder	Gravel	Conglomerate
−7	128	Cobble		
−6	64	Pebble		
−5	32			
−4	16			
−3	8			
−2	4	Granule		
−1	2	Very coarse	Sand	Sandstone
0	1	Coarse		
1	0.5	Medium		
2	0.25	Fine		
3	0.125	Very fine		
4	0.0625	Coarse	Silt	Siltstone
5	0.031	Medium		
6	0.015	Fine		
7	0.0078	Very fine		
8	< 0.0039	Clay	Clay	Claystone

Source: Modified after Wentworth, C.K., 1922. A scale of grade and class terms for clastic sediments. The Journal of Geology, 30, 377–392.

depositional environments play a significant role in sedimentary texture (size and sorting) and the structure of sandstones. Also, the mineralogical composition of sandstones is an indication of the source rock (provenance) properties, climate, and tectonic setting.

2.2.1 Components of sandstones

Like other sedimentary rocks, sandstones are ideally a complex consisting of four main elements, including framework (minerals and rock fragments), matrix, cement, and pore spaces (Fig. 2–1).

The framework of sandstones consists of different types of minerals and rock fragments, with specific textural properties, which have been derived from the destruction of rocks in the source area by weathering and erosion, and transportation by sedimentary processes (mostly water and wind), and finally deposition under the influence of sedimentary environment condition.

The matrix of sandstones contains fine sedimentary particles between the framework grains. It is typically composed of clay minerals, but also

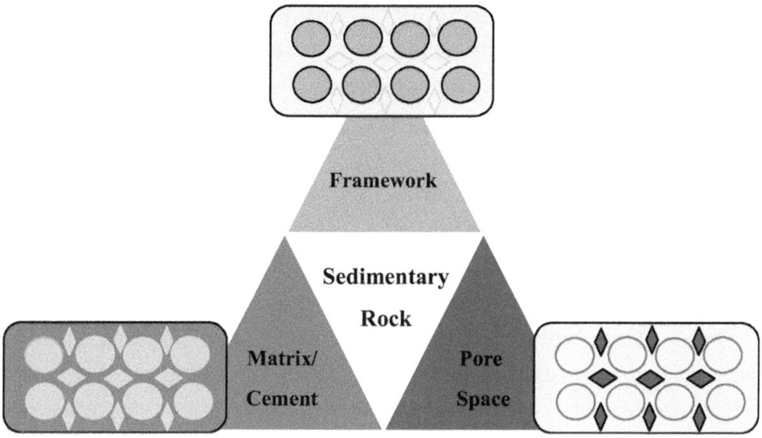

Figure 2–1 Schematic diagram showing three main components of sedimentary rocks.

includes very fine particles (in the size of silt and clay) of other minerals such as quartz and feldspar. Cement contains all crystalline materials within the pore spaces of the rock that originate from the infiltrated saturated fluids under the physicochemical conditions of porous media. Unlike the matrix, cement is developed within the rock after deposition and during different stages of diagenesis.

The mineralogical and chemical composition of sandstones is directly related to factors such as provenance (source rock composition and texture), climate, and tectonic setting (Dickinson and Suczek, 1979; Bhatia, 1983; Nesbitt et al., 1996; Zhiming et al., 2003). Many researchers in recent years used mineralogy and geochemical parameters as implications for the provenance, tectonic setting, and weathering history of sandstones (e.g., Sun et al., 2012; Zaid, 2015; Shi et al., 2016; Lin et al., 2019; Ojo et al., 2021). Therefore, due to the intimate relationship of mineralogy with the provenance and tectonic setting of sandstones, and also the effect of the sedimentary environment on their sedimentary texture and structure, the study of the sandstones based on their composition and texture helps to analyze the tectonic-sedimentary conditions of their formation, based on which the depositional environment and reservoir characteristics of these rocks can be investigated. Climate conditions through temperature and precipitation intensity have the main control on the intensity of weathering in the source area and, the resultant composition of the weathering products. For example, feldspars as the most important chemically unstable components in clastic rocks (Folk, 1980; Zuffa, 1985;

Pettijohn et al., 1987; Tucker, 1991) are highly sensitive to chemical weathering, especially in humid climate conditions, where they are altered to clay minerals. In the following section, these characteristics are discussed, separately.

2.2.1.1 Mineralogical composition

Quartz and feldspars, due to the abundance in the rocks of the Earth's crust, as well as the physical and chemical resistance, make up a significant percentage of sandstones mineralogy. The quartz grains, which are the most abundant particles (35%–50%) within the sediments, depending on the optical characteristics under the microscope, such as extinction (direct or undulate), the number of crystals (monocrystalline or polycrystalline), and the presence or absence of inclusions provide specific information of the source rocks (igneous, sedimentary, and metamorphic) (Adams et al., 1984).

Feldspar minerals include alkali feldspars and plagioclase which are derived from the weathering of igneous and metamorphic rocks. The less abundance (5%–15%) of these minerals with respect to quartz in sandstones is attributed to their lower physical and chemical resistance. Also, the chemical stability of plagioclase minerals including sodium (Albite: $NaAlSi_3O_8$) and calcium (Anortite: $CaAl_2Si_2O_8$) feldspars are less than potassium feldspars ($KAlSi_3O_8$), such as orthoclase and microcline. The presence of feldspars in sandstones depends on various factors including source rock properties, climate condition, and the effect of tectonic setting and transportation. A significant percentage of feldspars in sediments could be related to the arid climate, rapid deposition related to a tectonically active setting, and short transportation (e.g., Amini, 2011).

Rock fragments, which are broken components of sandstones, actually provide direct evidence of the source rock characteristics, from which the sediments are derived. Rock fragments are igneous, metamorphic, and sedimentary types that usually form coarse grains of sandstones.

Accessory minerals include mica (biotite and muscovite), heavy minerals, and clays that normally form a small percentage of components in sandstones. Heavy minerals that have a higher specific gravity (>2.85 g/cm^3) than the main minerals (quartz and feldspar), according to the optical characteristics under the microscope, are classified into opaque and non-opaque (light) types. The common light-heavy minerals include Garnet, Epidote, Staurolite, Apatite, Zircon, Rutile, and Tourmaline, and opaque heavy minerals are Hematite, Magnetite, and Ilmenite. Some of these minerals, such as Garnet, Epidote, and Staurolite are unique to metamorphic source rocks. Heavy minerals, due to their high density and smaller grain size, have different hydrodynamic behaviors with respect to the light minerals such as quartz and feldspars (Komar, 2007). Different factors affect the accumulation of heavy minerals from the source to the sedimentary environments. These factors include weathering at different stages between the source rocks and sedimentary environments, mechanical abrasion during transportation, physical sorting, and diagenetic processes during burial (Morton and Hallsworth, 1994, 1999). Zhu et al. (2017) used heavy mineral compositions of Cenozoic sandstones for the study of tectonic evolution and provenance of the SW Qaidam basin in the northern Tibetan Plateau.

Clay minerals are hydrous aluminosilicates belonging to the phyllosilicate group. Five dominant groups of clay minerals in sandstones are kaolinite, illite, chlorite, smectite, and mixed-layer types (Worden and Burley, 2003). Clay minerals can be incorporated into sandstones through depositional and post-depositional processes. Depositional clays include detrital mud matrix, mud intraclasts, flocculated mud, clay-rich rock fragments, and clay coats on sand grains. Post-depositional mechanisms include bioturbation, clay infiltration, and reactions during different stages of diagenesis (exogenetic, mesogenic, and telogenetic) (Worden and Burley, 2003).

In addition to the main and accessory minerals, some components are rare according to the environmental conditions seen in sandstones. Some of these, which are mainly diagenetic and secondary in nature, include pyrite, glauconite, phosphate, and organic matter.

2.2.1.2 Chemical composition

The chemical composition of sandstones is closely related to their mineralogical composition. The main silica-based minerals are mostly derived from igneous and metamorphic rocks. In addition, oxides share a minor percentage of minerals in these rocks. Table 2—2 summarizes the common minerals of sandstones with their chemical composition. The chemical composition and thus the mineralogy of sandstones can change significantly under the influence of diagenesis. During diagenetic processes, unstable minerals are eliminated entirely or partially from the rock, and some new minerals are added to the rock system (Pettijohn et al., 1972).

2.2.2 Sedimentary texture

Sedimentary texture describes three parameters related to the sediment particles including grain size, grain shape, and fabric.

Table 2–2 Common minerals of sandstones with their chemical composition.

Group	Subgroup	Mineral	Chemical formula
Silicates	Quartz	Quartz	SiO_2
	Feldspar	K-Feldspars (Orthoclase, microcline, sanidine)	$KAlSi_3O_8$
		Na-Plagioclase (Albite)	$NaAlSi_3O_8$
		Ca-Plagioclase (Anortite)	$CaAl_2Si_2O_8$
	Clay minerals	Kaolinite	$Al_2Si_2O_5(OH)_4$
		Illite	$(K, H_3O)(Al, Mg, Fe)_2(Si, Al)_4O_{10}[(OH)_2, (H_2O)]$
		Chlorite	$(Mg, Fe)_6(Si, Al)_4O_{10}(OH)_8$
		Montmorillonite/smectite	$(Na, Ca)_{0.33}(Al, Mg)_2Si_4O_{10}\ (OH)_2 \cdot nH_2O$
	Mica	Muscovite	$KAl_2(AlSi_3)O_{10}(OH)_2$
		Biotite	$K(Mg, Fe)_3(AlSi_3)O_{10}(OH)_2$
	Heavy minerals	Amphibole	$Ca_2(Mg, Fe)5Si_8O_{22}(OH)$
		Pyroxene	$Ca(Mg, Fe)Si_2O_6$
		Olivine	$(Mg, Fe)_2SiO_4$
		Garnet	$(Ca, Mg, Fe, Mn)_3Al_2Si_3O_{12}$
		Epidote	$Ca_2(Al, Fe)_3O(SiO_4)(Si_2O_7)(OH)$
		Zircon	$ZrSiO_4$
		Tourmaline	$(Na, Ca)(Al, Li, Mg)_{3-}(Al, Fe, Mn)_6(Si_6O_{18})(BO_3)_3(OH)_4$
		Sphene	$CaTiSiO_5$
Oxides		Magnetite	Fe_3O_4
		Ilmenite	$(Fe, Mg, Mn)TiO_3$
		Hematite	Fe_2O_3

2.2.2.1 Grain size

Is an important textural parameter to analyze the depositional conditions in terms of the energy and the processes dominating the sedimentary environment. The most commonly used scale for the grains size of sediment is from Wentworth (1922) which considers the large diameter of grains in millimeters (Table 2–1). For simplicity, and in a practical view, the millimeter scale of Wentworth was converted to the logarithmic scale of Krumbein (1934), known as the phi (φ) scale (Eq. 2.1).

$$\varphi = -\log_2 d \tag{2.1}$$

In this equation, d is grain diameter (mm).

Therefore, sandstones known as medium-sized sedimentary rocks, have a wide range of grain sizes, varying from very coarse to very fine sand (Table 2–1). Changes in the physical condition of the sedimentary environment cause changes in energy reflected in the grain size of the sediment. A uniform distribution of grain size indicates the continuity in sedimentary environment conditions with no or insignificant change in energy, which resulted from sediments with good sorting. In contrast, the sediments with wide variations in grain size or poor sorting show variability in energy or special conditions (e.g., rapid sedimentation) of the environment.

From a sedimentology point of view, analysis of grain size of sediments based on statistical and mathematical calculations gives useful and practical information about their depositional environment and the processes governing their sedimentation. These calculations are accomplished based on the data derived from granulometry analysis of sediments through different methods and techniques including direct measurement by caliper for coarse-grained sediments (granule) and sieve for medium-grained sediments (sand) and pipette and hydrometer for fine-grained sediments (silt and clay). For more details on these methods and techniques see Folk (1980) and Tucker (1988). Data obtained from the granulometric analysis are used for the calculation of some statistical parameters related to the grain size of the sediment such as mean grain size, standard deviation, skewness, and kurtosis. For calculation of these parameters, proposed two graphical and computational (arithmetic) methods. The formula for both methods is outlined in Table 2–3. Analysis of sediment based on the grain size parameters provides significant information related to the sedimentary processes and the physical condition dominating at the time of sediment deposition, by which the sedimentary environment can be interpreted.

Table 2–.3 Graphical and arithmetic formula for calculation of the main statistical parameters for grains size analysis of a sediment (after Folk, 1980).

Parameter	Graphical	Arithmetic	Description limits
Mean (Mz)	$\frac{\varphi16+\varphi50+\varphi84}{3}$	$\overline{x}=\sum\frac{fm}{n}$	Grain size class for sand: 2–1 mm (very coarse sand) 1–0.5 mm (coarse sand) 0.5–0.25 (medium sand) 0.25–0.125 (fine sand) 0.125–0.0625 (very fine sand)
Standard Deviation (σ_I)	$\frac{\varphi84-\varphi16}{4}+\frac{\varphi95-\varphi5}{6.6}$	$\sigma=\sqrt{\frac{\sum f(m-\overline{x})^2}{n}}$	Under 0.35φ (very well sorted) 0.35–0.5φ (well sorted) 0.5–0.71φ (moderately well sorted) 0.71–1φ (moderately sorted) 1–2φ (poorly sorted) 2–4φ (very poorly sorted) over 4φ (extremely poorly sorted)
Skewness (SK_I)	$\frac{\varphi16+\varphi84-2\varphi50}{2(\varphi84-\varphi16)}+\frac{\varphi5+\varphi95-2\varphi50}{2(\varphi95-\varphi5)}$	$SK=\frac{\sum f(m-\overline{x})^3}{100\sigma^3}$	+ 1 to + 0.3 (strongly fine-skewed) + 0.3 to + 0.1 (fine -skewed) + 0.1 to − 0.1 (near symmetrical) − 0.1 to − 0.3 (coarse skewed) − 0.3 to − 1 (strongly coarse skewed)
Kurtosis (K_G)	$\frac{\varphi95-\varphi5}{2.44(\varphi75-\varphi25)}$	$K=\frac{\sum f(m-\overline{x})^4}{100\sigma^4}$	Under 0.67 (very platykurtic) 0.67–0.9 (platykurtic) 0.9–1.11 (mesokurtic) 1.11–1.5 (leptokurtic) 1.5–3 (very leptokurtic) Over 3 (extremely leptokurtic)

f, Frequency for each classes of grain size in percent or weight; *m*, median value in a class of grain size; *n*, sum of total sediment in percent or weight; $\overline{x}$: mean of grain size.

2.2.2.2 Grain shape

Can be expressed based on parameters such as Form, Roundness, and Sphericity. The form describes the relationship between the three dimensions of a sedimentary grain (L = large, I = medium, S = short), which based on Zingg's (1935) scheme is classified into equant, oblate, bladed, and elongate (Fig. 2–2). Roundness is a criterion of the sharpness of the corners of a grain. Sphericity is a measure of the similarity between a grain and a sphere. In general, the degree of grain roundness and sphericity shows a direct relationship with the distance of transport from the source rock. But, the effect of transportation distance based on the grain composition, size, and also the nature of bedrock is different. As grains with coarse size and unstable mineralogy, and in contact with a rough bedrock experience rapid changes in shape, even at short distances. In addition, reworked particles originated from pre-existing sedimentary rocks because experience the second sedimentation cycle, and show a higher degree of sphericity and especially roundness. In fact, among different parameters of the shape, roundness is the main factor giving useful information on depositional texture related to the sedimentary environment. Fig. 2–3 show different classes of Powers (1953) scale for a description of grain roundness and sphericity. According to this scale, grains with similar sphericity can show different degrees of roundness, meanwhile, grains with similar roundness can have different degrees of sphericity.

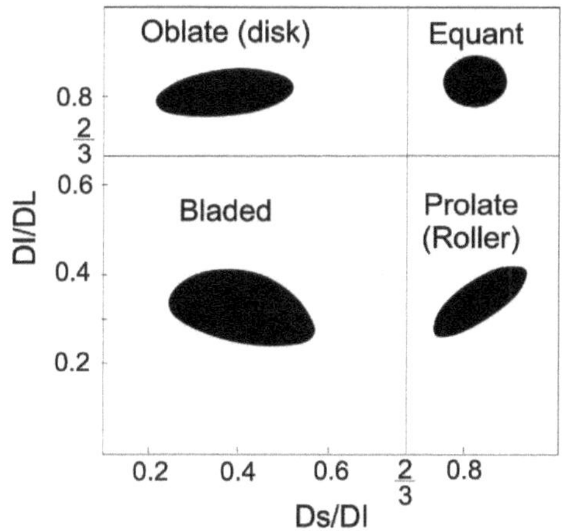

Figure 2–2 Zingg (1935) scheme for classification of sediment grain. *DL*, large diameter; *DI*, medium diameter; *Ds*, Short diameter (Sam Boggs, 2009). *Form modified after Boggs Jr. S., 2009. Petrology of Sedimentary Rocks. Cambridge University Press, second ed. 612 pp.*

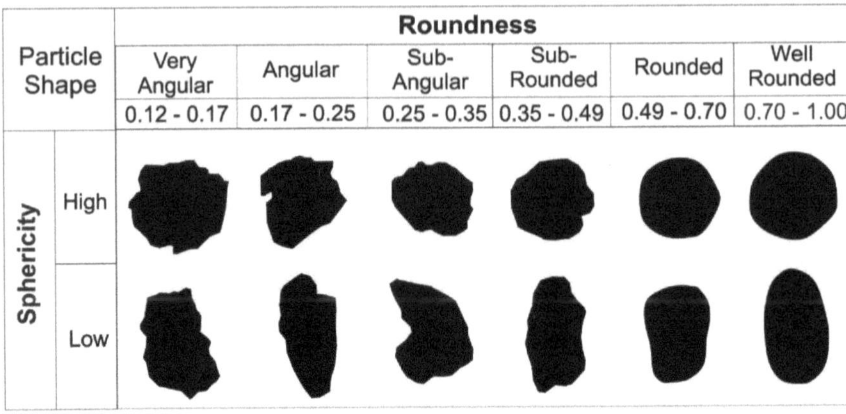

Figure 2–3 Different classes for a description of grain shape related to roundness and sphericity. *Modified after Powers, M.C. (1953). A new roundness scale for sedimentary particles. Journal of Sedimentary Petrology 23, 117–119.*

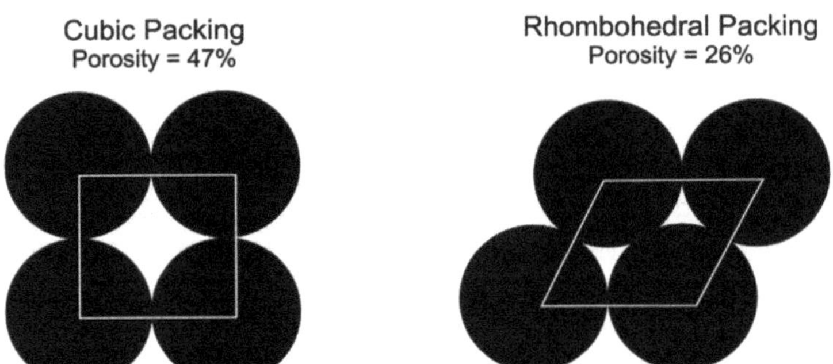

Figure 2–4 Two extremes of grains packing in sediment including open (cubic) and closed (rhombohedral) spacing.

2.2.2.3 Fabric

The orientation and packing of sediment grains are named fabric. Orientation refers to the arrangement of sediment grains in a certain direction relative to fluid flow. A specific type of orientation is named imbrication in which sediment grains show an arrangement with their inclination upstream. Packing refers to the spacing of the sediment grains and varies between two extremes; cubic and rhombohedral. In cubic packing, grains have a loose and open spacing with the maximum porosity (48%). In contrast, grains with rhombohedral packing have a tight and more compacted spacing with the minimum porosity (26%) (Fig. 2–4).

2.2.3 Textural maturity

Textural maturity provides information about the condition under which the sediment has been formed. The criteria for investigation of textural maturity of sediment respectively are clay content (a measure of energy), sorting (a measure of sedimentary environment and process), and roundness (a measure of transportation history). Four stages of textural maturity based on these criteria are immature, submature, mature, and super mature (Folk, 1951) (Table 2–4). Various sedimentary environments show different degrees of textural maturity. As, sediments with low degrees of maturity such as floodplain, lagoon, and alluvial fan belongs to weak action of currents or rapid deposition in environments. In contrast, mature and super-mature sediments are related to the intense and continuous action of currents on sediment, which are observed in beach and desert environments. Fig. 2–5 shows sample photomicrographs of immature and mature sandstone samples.

Table 2–4 The main sedimentary parameters for evaluation of textural maturity of sandstones.

Textural maturity	Clay content (%)	Sorting	Roundness
Immature	>5	–	–
Submature	<5	Poor	–
Mature		Well	Subangular/angular
Super-mature			Sub-rounded/rounded

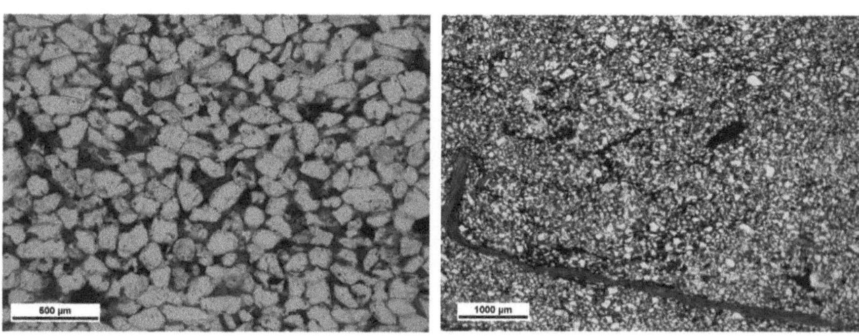

Figure 2–5 Photomicrographs of sandstone samples with mature (left) and immature (right) texture from the lower Cretaceous Gadvan Formation, southwest of Iran. The mature sandstone is characterized by low matrix (<5%), well-sorted, and rounded grains. The immature sandstone has a high percent of matrix (>5%), poorly sorted, and subangular grains.

Therefore, textural maturity is mainly dependent on the action of mechanical energy in winnowing, sorting, and abrading sediment in a sedimentary environment.

2.2.4 Sandstones classification

Sandstones are described and named based on the changes and diversity in their textural and mineralogical characteristics. The most important textural parameter is grain size, which is a measure of environmental energy. Udden and Wentworth (1922) have considered this criterion in their textural classification, and they have divided sandstones into five-grain size classes, including very fine, fine, medium, coarse, and very coarse (Table 2–1). In this classification, coarse grains represent more energetic environmental conditions than that fine grains. Since sandstones are not necessarily found in nature in a certain size and can be a combination of different sizes or mixed with gravel and silt and clay particles, Folk (1980) introduced a more practical classification to describe the clastic sediments. For this purpose, he proposed two triangular diagrams. In a large diagram, gravel, sand, and mud are the main components, and a small diagram (without gravel) encompasses sand, silt, and clay as the principal components (Fig. 2–6). According to this figure, sandy sediments include six

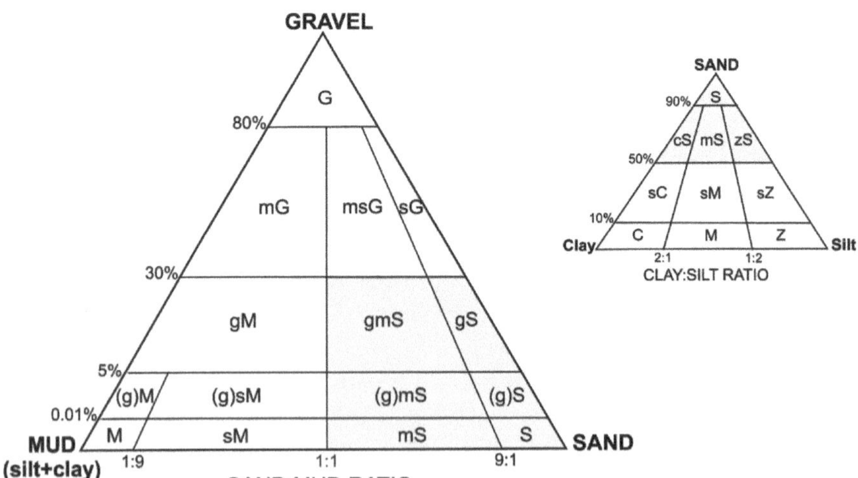

Figure 2–6 Triangular diagrams for classification of clastic sediments with Gravel (left) and without Gravel (right) content (). Sand groups have been highlighted on the diagrams. G, gravel; S, sand; M, mud; S, sand; Z, silt; C, clay; g, gravelly; (g), slightly gravelly; s, sandy; m, muddy; z, silty; c, clayey.

major groups of the large diagram, with less than 30% gravel and a ratio of sand to the mud of more than 1, including sand (S), gravelly sand (gS), muddy sand (mS), gravelly muddy sand (gmS), slightly gravelly sand ((g)S), and slightly gravelly muddy sand ((g)mS). Also, four major groups of small diagrams with more than 50% sand show sandy sediments, which are sand (S), silty sand (zS), muddy sand (mS), and clayey sand (cS). Fig. 2–7 demonstrates some core photographs of clastic sediments from Ojageshlagh and Zeivar formations in Moghan Basin, northwest Iran. These formations show a heterolithic lithology varying from shale and fine-grained sandstone to coarse and gravelly sandstone.

Figure 2–7 Core photographs of clastic sediments with different textures (grain size) from Ojageshlagh and Zeivar formations of Moghan area, NW Iran. (A) Shale (mixture of silt and clay with lamination), (B) Fine sandstone, (C) Medium sandstone, (D) Coarse sandstone, (E) Gravely sandstone, (F) Conglomerate. A, C and F belong to Ojageshlagh tight Formation, and B, D, and E belong to Zeivar tight Formation. The length of the yellow line for each photo is 10 cm.

On the other hand, the description of sandstones based on mineralogy provides useful information on their provenance, tectonic setting, and transportation history. The presence and frequency of different components (minerals and rock fragments) in sediment is a function of their availability and physical durability and chemical stability (Folk, 1980). The availability of different minerals is mainly dependent on the rocks in the source area, from which the sediment originated (sediment provenance). Physical durability defines the resistance of minerals against abrasion and fracturing, which is mostly controlled by their crystalline structure. Chemical stability indicates the sensitivity of minerals with respect to chemical reactions. Quartz due to hardness and lack of cleavage has high durability and chemical stability. In contrast, feldspars that are softer than quartz, and have cleavage, show low durability. These minerals also are more sensitive to chemical fluids than quartz and they are dissolved or altered to clay minerals. Generally, sediments with a significant percent of stable components such as quartz and chert have a high degree of mineralogical maturity compared to the sediments rich in feldspars and unstable rock fragments (e.g., volcanic, shale, carbonate, and metamorphic lithics). Tectonic and climatic factors have the main control over the mineralogical maturity of sandstones. As areas with tectonic stability are characterized by low topography, low rate of subsidence, and as a result, low rate of sediment deposition. In these areas, time is sufficient for the effective action of chemical weathering on feldspars, especially in a humid climate. Therefore, the percent of chemically unstable components such as feldspars in the sediment is decreased, which results in mature sediment. Transportation and increasing the distance from the source have also an effective impact on abrasion and decrease the feldspars and unstable rock fragments (e.g., shale, slate, and phyllite) in the sediment.

One of the common classifications of sandstones based on mineralogy is the Folk triangle (Folk, 1980), which consists of three main components of quartz, feldspar, and rock fragments (lithics). Sandstones in this scheme, based on the abundance of these components, are classified into three main groups, including quartz arenite, arkose, and litharenite, as well as their intermediate types (Fig. 2–8). Furthermore, in another classification from Pettijohn et al. (1987), in addition to the three components of the Folk triangle, the matrix is used as the fourth component. In this scheme, sandstones are classified into arenites with less than 15% matrix, and graywackes with more than 15% matrix (Fig. 2–9). Generally, mineralogically immature sandstones (i.e., arkose and litharenites) when exposed to

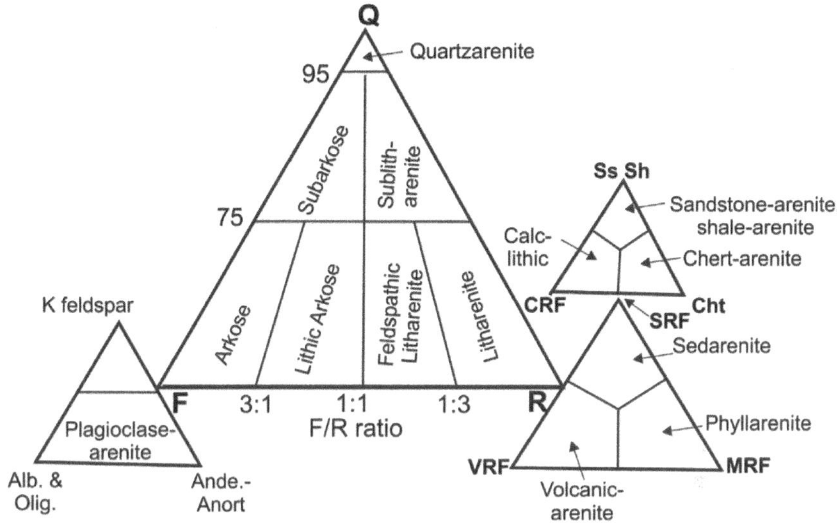

Figure 2–8 Classification of sandstones based on three main components (Q, quartz; F, feldspar; and R, rock fragments) on the Folk diagram (Folk, 1980).

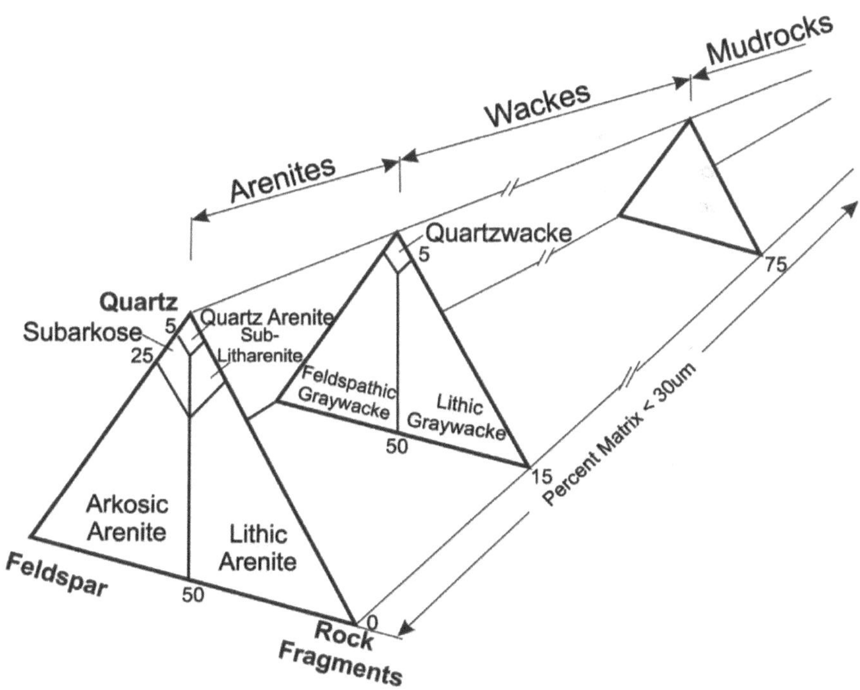

Figure 2–9 Classification of sandstones based on the triangles of Pettijohn et al. (1987). In this classification, arenites (matrix <15%) and wackes (matrix >15%) are differentiated.

suitable conditions (humid climate and stable tectonic setting), can gradually be converted to submature (sub-arkose and sub-litharenites), and finally mature sandstones (quartz arenite).

2.2.5 Reservoir quality

Porosity and permeability are considered the most important physical properties of reservoir rocks (Tiab and Donaldson, 2004). Sandstones in terms of their capacity for fluid storage (porosity) and capability for fluid flow (permeability) include a significant part of hydrocarbon reservoir rocks in the world. Porosity is the ratio of pore spaces of rock to its total volume that is commonly expressed by the phi (φ) symbol in percent (Eq. 2.2). For example, in rock with 25% porosity, pore spaces share a quarter of the total volume of rock.

$$\Phi = \frac{\text{the volume of pore spaces}}{\text{the total volume of rock}} \tag{2.2}$$

$$= \frac{\text{the total volume of the rock} - \text{volume of solid rock}}{\text{the total volume of rock}} \times 100$$

In addition, a part of total porosity that includes interconnected pore spaces of the rock is named effective porosity. In fact, for reservoir engineers, effective porosity due to its essential role in fluid flow within the reservoir is more important than total porosity.

Permeability which is a measure of the rock's ability to transmit fluid in a porous medium is determined by Darcy's law according to Eq. 2.3. One Darcy describes the permeability in which a fluid with a viscosity of one centipoise and a pressure gradient of one atmosphere per centimeter, flows at one centimeter per second. Therefore, Darcy is the unit used for the expression of permeability values. However, because most reservoirs are characterized by the permeability of less than one Darcy, it is expressed in terms of millidarcy (md), which is one-thousandth of a Darcy.

$$Q = \frac{KA\Delta P}{\mu . L} \tag{2.3}$$

In this equation, Q is the rate of flow (cm^3/s), ΔP is the pressure gradient (bar), A is the cross-sectional area of the rock (cm^2), μ is the fluid viscosity (centipoises), L is the length of pressure gradient and k is permeability (md).

Porosity and permeability in clastic rocks such as sandstones are mainly controlled by depositional environment conditions that have the main control on textural attributes (e.g., grain size, grain shape, sorting, fabric) (Tiab and Donaldson, 2004). In fact, the effect of these conditions is recorded in depositional facies of sandstones. In addition, post-depositional processes (diagenesis) just after deposition can affect the reservoir characteristics of the sandstones. The effect of these parameters on porosity and permeability is described below.

2.2.5.1 Depositional facies

Depositional facies of sandstones in terms of textural characteristics such as grain size, shape, sorting, and fabric control their pore system properties and reservoir quality, as follows.

2.2.5.1.1 Grain size

The relationship between porosity and grain size is different for sand sediments and their equivalent rocks (sandstones). In unconsolidated sands, porosity is more dependent on sorting than grain size. But permeability is a function of both grain size and sorting (Frederic and Park, 1986) (Fig. 2–10). Generally, fine-grained sediments show higher porosity than coarse-grained ones. This is attributed to the grain shape, low roundness, and sphericity of

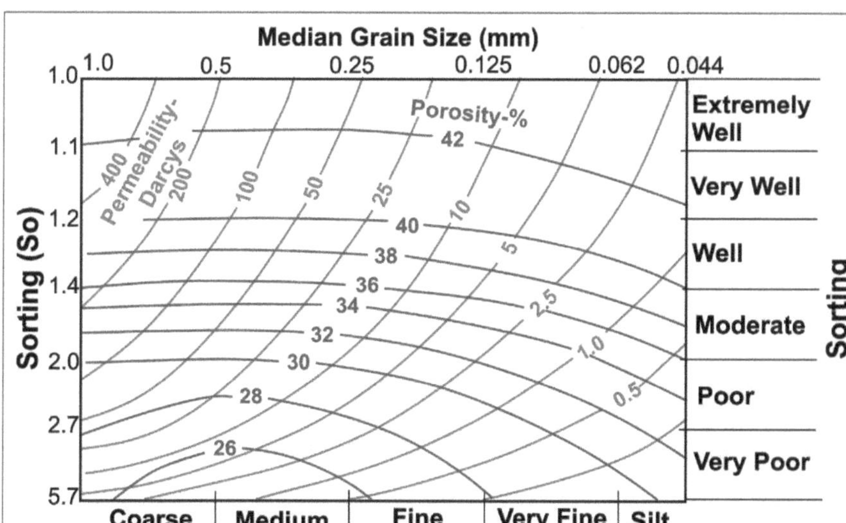

Figure 2–10 Porosity and permeability of artificially mixed and wet-packed natural sands from Brazos River, Texas. S_o is the sorting coefficient of Trask. *Modified by Beard and Weyl (1973).*

fine-grained sediments, which causes an open spacing between grains (open packing) and thus high porosity. In contrast, fine-grained sandstones tend to lose more volume of porosity during the burial, than they are compared with coarse-grained sandstones. Permeability shows a direct relationship with grain size; the coarser grain size, the higher permeability. This is attributed to larger pore throat sizes of coarse grains. It can be concluded that the grain size of sandstones has the main effect on their effective porosity and permeability.

2.2.5.1.2 Sorting

The relationship between this parameter with porosity and permeability is straightforward. Sediments with uniform grain size (well sorting) have higher porosity and permeability than poorly sorted sediments.

2.2.5.1.3 Grain shape

Sphericity and roundness are two important parameters of the grain shape that affect the porosity and permeability by controlling the degree of compaction and packing (Tiab and Donaldson, 2004). In general, sediment grains with a high degree of roundness and sphericity have a closed packing with low values of porosity and permeability.

2.2.5.1.4 Fabric (packing and orientation)

Orientation has the main effect on fluid flow and permeability in a direction that the sediment grains have been oriented. Packing as previously mentioned, controls the volume of pore spaces between grains, changing from closed (orthorhombic with 26% porosity) to open (cubic with 48% porosity) packing.

2.2.5.2 Diagenesis

Diagenesis involves all post-depositional processes that affect sediment just after deposition to before the onset of metamorphism (Selley, 2000). The study of the diagenetic history of sandstones helps in understanding the evolution of porosity and permeability in these rocks as potential oil and gas reservoirs (Wolela and Gierlowski-Kordesch, 2007). Some diagenetic processes such as compaction and cementation, lead to the conversion of sediment to a sedimentary rock (lithification), which is mainly accompanied by a reduction in pore volume or porosity removal. In contrast, other diagenetic processes, such as dissolution and fracture, are involved in the development of porosity or change of pore types. Sandstone diagenesis is controlled by many factors, including depositional environments and local tectonics, sea-level changes, mineralogical composition, and burial history (Kim et al., 2007). The size and type of depositional systems are mainly controlled by tectonic settings and climate, which in turn have the main control on the composition and maturity of sandstones (Behrenbruch and Biniwale, 2005). Changes in sea level during the sediment deposition, in addition to depositional facies, reflect in stratal surfaces and developing unconformities that account for some subaerial diagenetic processes such as dissolution and cementation.

Compaction as mechanical and chemical occurs under the effect of overburden pressure during the shallow and deep burial of sediment and causes porosity reduction. Factors controlling the degree of compaction in sandstones are burial rate, composition, mechanical and petrophysical properties, temperature, pressure, and stress conditions (Mørk and Moen, 2007). Mechanical compaction that is identified by re-arrangement and breaking of grains and squeezing of ductile grains, results in a dense packing fabric in shallow burial (Boggs, 1992, 2006; Chester et al., 2004). Following the physical compaction and increasing the compaction degree in deep burial, chemical compaction as a result of pressure solution is initiated, in which material is removed by solution or diffusion from a discrete surface between grains contacts (Groshong, 1988). The features of chemical compaction vary from the initial stage of pressure solution and small scale as smooth and sutured contacts of detrital grains, to stylolite in a wide-scale and higher degree of compaction. Stylolites are considered intergranular serrated surfaces lined by insoluble materials of the enclosing rock (Tada and Siever, 1989). These surfaces that are perpendicular to the direction of maximum stress (Onasch, 1994), form bedding-parallel (resulting from overburden pressure), and not-bedding-parallel related to tectonic compression (Railbeck and Andrew, 1995). Therefore, chemical compaction due to its effect on greater packing, and also as a major source of silica for cementation through pressure solution, has a destructive effect on the reservoir quality of sandstones (Giles et al., 2000; Baron and Parnell, 2007).

Diagenetic cements have significant control on reservoir properties by reducing pore spaces (Blamey et al., 2014). Different cements, in addition to their effects on porosity and permeability, can alter the rock strength (Al-Tahini et al., 2006; Webster and Taylor, 2007). Three common cements found in sandstones are carbonates, silica, and clay minerals. Other cements of local importance and less abundance include anhydrite (and gypsum), celestite, halite, hematite, zeolite, barite, and feldspar (Selley, 2000).

The effect of quartz cementation on reservoir quality in sandstones has been considered by many researchers (e.g., Bjørlykke et al., 1986; McBride, 1989; Worden and Burley, 2003; Peltonen et al., 2009). The silica for quartz cement can be provided from different sources including clay mineral transformation (e.g., smectite to illite) during burial diagenesis of shales, alteration of detrital feldspars (i.e., Na-plagioclase and K-feldspar) (Metwally and Chesnokov, 2012), and pressure solution of detrital quartz (Renard et al., 2000; Cyziene et al., 2006). Carbonate cements in addition to porosity may have a significant influence on fluid flow (e.g., McBride, 1989; Morad, 1998; Dutton et al., 2002). Clay cements depending on their origin, the positions that they occupy within the pore system, and also their structure and arrangement, have a different effect on reservoir quality (Fig. 2−11). Kaolinite, smectite, illite, and chlorite are considered the main authigenic clay minerals in the sandstone reservoirs. Generally, clays with a tangential arrangement on grain surfaces, have less effect on permeability than clays with a perpendicular arrangement (Bushell, 1986). Kaolinite as a pore-filling clay has the main effect on porosity reduction. In contrast, pore-bridging clays such as illite have the main impact on pore throats and reduce permeability. In addition, the presence of chlorite as a grain coating and pore-lining clay prevents the development of silica as an overgrowth cement around quartz grains (Ehrenberg, 1993; Storvoll et al., 2002).

2.3 Comparison of depositional facies, environment, and diagenetic characteristics of tight sandstones with conventional sandstones

What are the factors to distinguish tight sandstone reservoirs from their equivalent conventional reservoirs? The answer to this question can be different according to the geological, reservoir, and production aspects of these reservoirs. Based on the results of studies of various tight sandstone reservoirs around the world, it can be found that these reservoirs are not limited to specific geological settings and certain times. As Coleman (2008) provided a discussion on challenges related to the discovery and development of tight gas sandstone reservoirs by an overview of representative examples of different ages from Paleozoic to Cenozoic and different basinal settings from continental fluvial to deep marine in many US basins. General characteristics of these reservoirs have been summarized in Table 2−5. Some results are derived from the

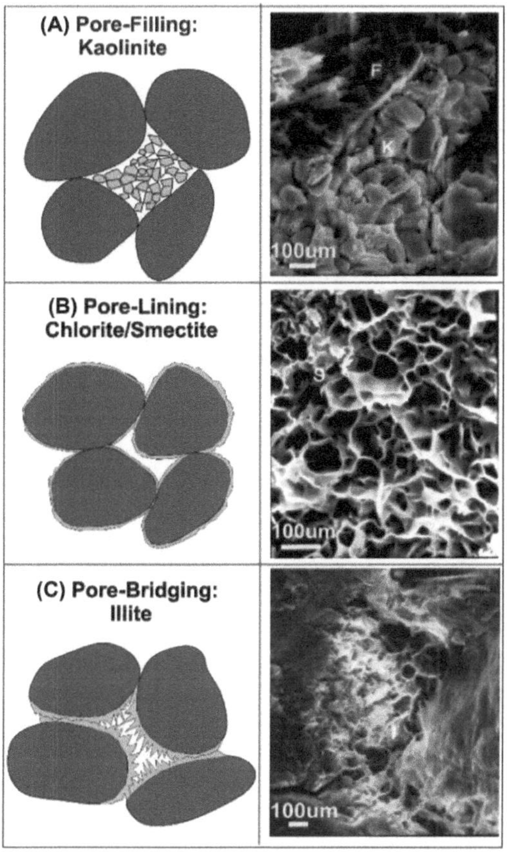

Figure 2–11 A schematic view of the effect of authigenic clays as pore-filling (Kaolinite), pore-lining (Chlorite/Smectite), and pore-bridging (Illite) on pore system properties of reservoir rocks (Neasham, 1977). *Modified after Neasham, J.W., 1977. The morphology of dispersed clay in sandstone reservoirs and its effect on sandstone shaliness, pore space and fluid flow properties. In: SPE 6858 Annual Fall Technical Conference and Exhibition. SEM photomicrographs are from the tight sandstones of the Whicher Range filed in Perth Basin of Western Australia.*

discussion as follows. (1) structural position does not guarantee economic gas production from tight sandstones, (2) local and regional variations in sandstone facies and burial-related diagenesis control reservoir quality and commercial gas production of these reservoirs, and (3) tight sandstones may be continuous reservoirs, but their reservoir properties are not necessarily continuous across the full extent of the reservoir. Moreover, a comprehensive discussion on correlations and differences between conventional and unconventional oil and gas reservoirs has

Table 2–5 Summary of tight-gas sandstone reservoirs with different ages and basinal settings in a number of US basins.

Basin	Group/formation	Year	Age	Depositional setting	Porosity (%)	Permeability (md)
Appalachian basin	Tuscarora sandstone	1978	Silurian	Shallow water mixed-marine deposits	0.5	0.05
Black Warrior basin	Pottsville sandstone	1978	Pennsylvanian	Marine-dominated deltaic complex	-	-
Gulf of Mexico basin	Cotton valley sandstone	1980	Jurassic	Marine-dominated deltaic complex	2–8	0.022–0.41
Appalachian basin	Oriskany sandstone	1982	Devonian	Shallow-water marine sandstone	0–12	0.10–29.6
Gulf of Mexico basin	Wilcox group sandstone	1984	Eocene	Shallow-water, muddy, marine-shelf	-	-
Arkoma basin	Jackfork group sandstone	1996	Pennsylvanian	Deep-water turbidite and debrite	-	-
Greater green river basin	Frontier sandstone	1999	Cretaceous	Deltaic sandstone and shale complex	8–12	0.01–0.04
Eastern green river basin	Almond sandstone	2003	Cretaceous	Marine barrier-bar and tidal-coastal plain	10–11	-

Source: Data and information adapted from Coleman, Jr. J.L., 2008. Tight-gas sandstone reservoirs: 25 years of searching for "the answer," in S. P. Cumella, K. W. Shanley, and W.K. Camp, (eds.), Understanding, exploring, and developing tight-gas sands. 2005, Vail Hedberg Conference: AAPG Hedberg Series, no. 3, p. 221–250.

been provided by Pang et al. (2021). For this purpose, they focused on six representative basins of China, and also considered the progress of unconventional oil/gas exploration in North America and the distribution characteristics of oil/gas reservoirs in 1186 basins around the world. They discussed that the conventional and unconventional reservoirs show the same characteristics in terms of oil and gas sources, geological age, sedimentary basins, and petroleum systems. In contrast, they show differences in hydrocarbon compositions, spatial relations to source rocks, reservoir lithology and qualities, geological settings, and the mechanisms involved in their formation. In general, tight sandstones genetically show similar characteristics in terms of mineralogy, texture, and depositional setting to conventional sandstone reservoirs. This means tight gas sandstones were not always as tight as they are now (Coleman, 2008). Therefore, the main difference is thought to be related to the pore structure of tight sandstones that does not allow for economic production rate in tight reservoirs, as the word "tight" is equivalent to the "low permeability" in the literature of these reservoirs. Thus, it is essential to identify various factors that cause a sandstone to be tight. Diagenetic cements such as quartz, calcite and clays, complex pore network consisting of disconnected pore spaces and tortuous pore throats as well as high volume of irreducible water saturations are among the factors that limit gas production and fluid flow in tight sandstones, and are discussed in next chapters. Fig. 2–12 demonstrates a comparison of two unconventional (tight) sandstones from Perth basin of Western Australia and Moghan Basin of Iran with a conventional sandstone reservoir from Zagros basin of Iran on porosity-permeability cross plot. According to this figure, tight sandstones show wide range of porosity, but narrow range of permeability that means their pore system consists mostly of pores with limited and poor connection (isolated pores and micropores). In contrast, high values of porosity and permeability in conventional sandstone indicate a pore system with good connection of pore spaces. Also, two tight reservoirs show differences in permeability values on the plot, as Moghan sandstones generally are tighter than sandstone reservoirs of the Perth Basin. This is attributed to different mineralogy and different daigenetic history of theses reservoirs. Regarding to the evidences from core description and petrographic studies, sandstones of Moghan basin are mostly volcanic litharenites that have experienced high degree of compaction compared to the feldspathic arenites of Perth Basin.

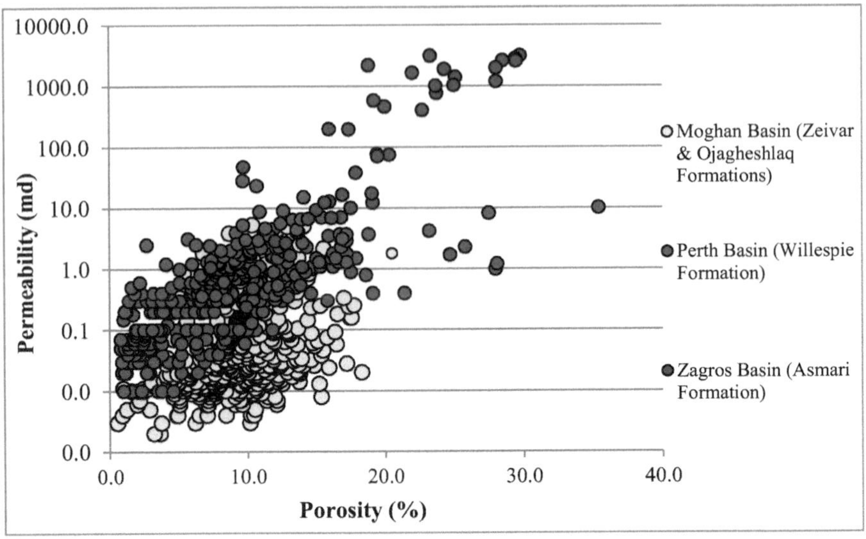

Figure 2–12 Cross plot of core porosity and permeability for unconventional (tight) sandstones from Moghan (NW Iran) and Perth basins (Western Australia) associated with conventional sandstones of Zagros Basin (SW Iran).

References

Adams, A.E., Mackenzie, W.S., Guliford, C., 1984. Atlas of Sedimentary Rocks under the Microscope, first ed. Routledge, p. 110.

Al-Tahini, A.M., Sondergeld, C.H., Rai, C.S., 2006. The Effect of Cementation on the Mechanical Properties of Sandstones. SPE 89069.

Amini, A., 2011. Use of feldspar grains in provenance determination and the study of transportation and depositional history, examples from central and NW Iran. Journal of Geopersia 1 (2), 11–24.

Baron, M., Parnell, J., 2007. Relationships between stylolites and cementation in sandstone reservoirs: examples from the north sea, UK and east greenland. Sedimentary Geology 194, 17–35.

Beard, D.C., Weyl, P.K., 1973. Influence of texture on porosity and permeability of unconsolidated sand. AAPG Bulletin 57 (2), 349–369.

Behrenbruch, P., Biniwale, T.S., 2005. Characterisation of clastic depositional environments and rock pore structures using the Carman-Kozeny equation: Australian sedimentary basins. Journal of Petroleum Science and Engineering 47, 175–196.

Bhatia, M.R., 1983. Plate-tectonics and geochemical composition of sandstones. Journal of Geology 91, 611–627. Available from: https://doi.org/10.1086/628815.

Bjørlykke, K., Aagaard, P., Dypvik, H., Hastings, D.S., Harper, A.S., 1986. Diagenesis and reservoir properties of jurassic sandstones from the haltenbanken area, offshore mid Norway. In: Spencer, A.M. (Ed.), Habitat of Hydrocarbons on the Norwegian Continental Shelf. Norwegian Petroleum Society, Graham & Trotman, pp. 275–286.

Blamey, N.J.F., Azmy, K., Brand, U., 2014. Provenance and burial history of cement in sandstones of the Northbrook Formation (Carboniferous), western Newfoundland, Canada: A geochemical investigation. Sedimentary Geology 299, 30–41.

Boggs Jr., S., 1992. Petrology of Sedimentary Rocks. Macmillan Publishing Company.

Boggs Jr., S., 2006. Principles of Sedimentology and Stratigraphy, fourth ed. Pearson Education, Inc, p. 676.
Boggs Jr., S., 2009. Petrology of Sedimentary Rocks, second ed. Cambridge University Press, p. 612.
Bushell, T.P., 1986. Reservoir geology of the morecambe field. In: Brooks, J., Goff, J.C., van Hoorn, B. (Eds.), Habitat of Palaeozoic Gas in N.W. Europe, 23. Special Publication of the Geological Society, London, pp. 189–208.
Chester, J.S., Lenz, S.C., Chester, F.M., Lang, R.A., 2004. Mechanism of compaction of quartz sand at diagenetic conditions. Earth and Planetary Science Letters 220, 435–451.
Coleman Jr. J.L., 2008. Tight-gas sandstone reservoirs: 25 years of searching for "the answer," in S. P. Cumella, K. W. Shanley, and W.K. Camp, (eds.), Understanding, exploring, and developing tight-gas sands. 2005, Vail Hedberg Conference: AAPG Hedberg Series, no. 3, pp. 221–250.
Cyziene, J., Molenaar, N., Sliaupa, S., 2006. Clay-induced pressure solutions as Si source for quartz cement in sandstones of the Cambrian Deimena Group. Geologija 53, 8–21.
Dickinson, W.R., Suczek, C.A., 1979. Plate tectonics and sandstone compositions. AAPG Bulletin 63 (12), 2164–2182.
Dutton, S.P., White, C.D., Willis, B.J., Novakovic, D., 2002. Calcite cement distribution and its effect on fluid flow in a deltaic sandstone, frontier formation, Wyoming. AAPG Bulletin 86 (12), 2007–2021.
Ehrenberg, S.N., 1993. Preservation of anomalously high porosity in deeply buried sandstones by grain-coating chlorite: examples from the norwegian continental shelf. American Association of Petroleum Geologists Bulletin 77, 1260–1286.
Folk, R.L., 1951. Stages of textural maturity in sedimentary rocks. Journal of Sedimentary Petrology 21, 127–130.
Folk, R.L., 1980. Petrology of Sedimentary Rocks. Hemphill Pub., Austin, TX.
Frederic, L., Park, W.C., 1986. Porosity reduction in sandstone by quartz overgrowth. AAPG Bulletin 70 (11), 1713–1728.
Giles, M.R., Indrelid, S.L., Beynon, G.V., Amthor, J., 2000. The origin of large-scale quartz cementation: evidence for large datasets and coupled heat-fluid mass transport modeling. In: Wordon, R.H., Morad, S. (Eds.), Quartz cementation in sandstones. In: 29. International Association of Sedimentologists Special Publication, pp. 21–38.
Groshong Jr., R.H., 1988. Low-temperature deformation mechanisms and their interpretation. Geological Society of America Bulletin 100, 1329–1360.
Kim, J.C., Lee, Y.I., Hisada, K., 2007. Depositional and compositional controls on sandstone diagenesis, the tetori group (Middle Jurassic-Early Cretaceous), central Japan. Sedimentary Geology 195, 183–202.
Komar, P.D., 2007. The entrainment, transport and sorting of heavy minerals by waves and currents. Developments in Sedimentology 58, 3–48.
Krumbein, W.C., 1934. Size frequency distributions of sediments. Journal of Sedimentary Petrology 4, 65–77.
Lin, N.H., Guo, Y., Wai, S.N., Tamehe, L.S., Wu, Z., Naing, N.M., et al., 2019. Sedimentology and geochemistry of middle eocene-lower oligocene sandstones from the western salin sub-Basin, the central Myanmar Basin: implications for provenance, source area weathering, paleo-oxidation and paleo-tectonic setting. Journal of Asian Earth Sciences 173, 314–335.
McBride, E.F., 1989. Quartz cement in sandstones: a review. Earth Science Reviews 26, 69–112.
Metwally, Y.M., Chesnokov, E.M., 2012. Clay mineral transformation as a major source for authigenic quartz in thermo-mature gas shale. Applied Clay Science 55, 138–150.

Morad, S., 1998. Carbonate cementation in sandstones: distribution patterns and geochemical evolution. The International Association of Sedimentologists. Available from: https://doi.org/10.1002/9781444304893.

Mørk, M.B.E., Moen, K., 2007. Compaction microstructures in quartz grains and quartz cement in deeply buried reservoir sandstones using combined petrography and EBSD analysis. Journal of Structural Geology 29, 1843–1854.

Morton, A.C., Hallsworth, C., 1994. Identifying provenance-specific features of detrital heavy mineral assemblages in sandstones. Sedimentary Geology 90, 241–256.

Morton, A.C., Hallsworth, C., 1999. Processes controlling the composition of heavy mineral assemblages in sandstones. Sedimentary Geology 124, 3–29.

Neasham, J.W., 1977. The morphology of dispersed clay in sandstone reservoirs and its effect on sandstone shaliness, pore space and fluid flow properties. In: SPE 6858 Annual Fall Technical Conference and Exhibition.

Nesbitt, H.W., Young, G.M., McLennan, S.M., Keays, R.R., 1996. Effects of chemical weathering and sorting on the petrogenesis of siliclastic sediments, with implications for provenance studies. Journal of Geology 104, 525–542.

Ojo, O.J., Suraju, A., Adepoju, S.A., Awe, A., Adeoye, M.O., 2021. Mineralogy and geochemistry of the sandstone facies of Campanian Lokoja formation in the southern bida Basin, Nigeria: implications for provenance and weathering history. Heliyon 7 (12), e08564.

Onasch, C.M., 1994. Assessing brittle volume-gain and pressure solution volume-loss processes in quartz arenite. Journal of Structural Geology 16, 519–530.

Pang, X., Shao, X., Li, M., Hu, T., Chen, Z., Zhang, K., et al., 2021. Correlation and difference between conventional and unconventional reservoirs and their unified genetic classification. Gondwana Research 97, 73–100.

Peltonen, C., Marcussen, Ø., Bjørlykke, K., Jahren, J., 2009. Clay mineral diagenesis and quartz cementation in mudstones: the effects of smectite to illite reaction on rock properties. Marine and Petroleum Geology 26, 887–898.

Pettijohn, F.J., Potter, P.E., Siever, R., 1987. Sand and Sandstone, third ed Springer-Verlag, New York.

Pettijohn, F.J., Potter, P.E., Siever, R., 1972. Sand and Sandstone. Springer Verlag, New York.

Powers, M.C., 1953. A new roundness scale for sedimentary particles. Journal of Sedimentary Petrology 23, 117–119.

Railbeck, L.B., Andrew, L.M., 1995. Tectonic stylolites in the 'undeformed' Cumberland Plateau of southern Tennessee. Journal of Structural Geology 17, 911–915.

Renard, F., Brosse, E., Gratier, J.P., 2000. The different processes involved in the mechanism of pressure solution in quartz-rich rocks and their interactions. Special Publications International Association of Sedimentologists 29, 67–78.

Selley, R.C., 2000. Applied Sedimentology. Cademic Press, London, p. 633.

Shi, G., Wang, H., Huang, G., Yang, S., Song, G., 2016. Provenance and tectonic setting of middle-upper Devonian sandstones in the Qinling Orogen (Shanyang area): new insights from geochemistry, heavy minerals and tourmaline chemistry. Tectonophysics 688, 11–25.

Storvoll, V., Bjørlykke, K., Karlsen, D., Saigal, G., 2002. Porosity preservation in reservoir sandstones due to grain-coating illite: a study of the Jurassic Garn Formation from the Kristin and Lavrans fields, offshore Mid-Norway. Marine and Petroleum Geology 19, 767–782.

Sun, L., Gui, H., Chen, S., 2012. Geochemistry of sandstones from the neoproterozoic Shijia formation, northern Anhui Province, China: implications for provenance, weathering and tectonic setting. Geochemistry 72 (3), 253–260.

Tada, R., Siever, R., 1989. Pressure solution during diagenesis. Annual Review of Earth and Planetary Sciences 17, 89–118.

Tiab, D., Donaldson, E.C., 2004. Petrophysics: Theory and Practice of Measuring Reservoir Rock and Fluid Transport Properties, second ed. Gulf Professional Publishing, p. 926.
Tucker, M.E., 1988. Techniques in Sedimentology. Blackwell Scientific Publications, Oxford; Boston, MA.
Tucker, M.E., 1991. Sedimentary Petrology: An introduction to the origin of sedimentary rocks. Blackwell Scientific Publication, London.
Webster, C.M., Taylor, P.G., 2007. Integrating quantitative and qualitative reservoir data in sand prediction studies: the combination of numerical and geological analysis. SPE 108586.
Wentworth, C.K., 1922. A scale of grade and class terms for clastic sediments. The Journal of Geology 30, 377–392.
Wolela, A.M., Gierlowski-Kordesch, E., 2007. Diagenetic history of fluvial and lacustrine sandstones of the Hartford Basin (Triassic–Jurassic), Newark Supergroup, USA. Sedimentary Geology 197 (1–2), 99–126.
Worden, R.H., Burley, S.D., 2003. Sandstone diagenesis: The evolution of sand to stone. In: Burley, S.D., Worden, R.H. (Eds.), Sandstone diagenesis. Recent and ancient. International Association of Sedimentologists Special Publication, 4, Blackwell, Oxford, pp. 3–44.
Zaid, S.M., 2015. Geochemistry of sandstones from the Pliocene Gabir Formation, north Marsa Alam, Red Sea, Egypt: implication for provenance, weathering and tectonic setting, Journal of African Earth Sciences, 102. pp. 1–17.
Zhiming, L., Jiajun, L., Ruizhong, H., Mingqin, H., Yuping, L., Chaoyang, L., 2003. Tectonic setting and nature of the provenance of sedimentary rocks in Lanping Mesozoic-Cenozoic Basin: evidence from geochemistry of sandstones. Chinese Journal of Geochemistry 22, 352–360.
Zhu, W., Wu, C., Wang, J., Zhou, T., Li, J., Zhang, C., et al., 2017. Heavy mineral compositions and zircon U-Pb ages of Cenozoic sandstones in the SW Qaidam basin, northern Tibetan Plateau: implications for provenance and tectonic setting. Journal of Asian Earth Sciences 146, 233–250.
Zingg, Th, 1935. Beitragezur Schotteranalyse. Schweizer. Mineralog. u. Petrog. Mitt. 15, 39–140.
Zuffa, G.G., 1985. Optical analyses of arenites: influence of methodology on compositional results. In: Zuffa G.G. (ed) Provenance of Arenites. NATO ASI Series D, Reidel Dordrecht; pp. 165–189.

Further reading

Arastoopout, H., Chen, S.T., 1987. Analysis of flow of gas and water in a fractured and non-fractured low-permeability reservoir under production: 1987 Society of Petroleum Engineers Annual Technical Conference, Dallas, Texas, September 27–30, 1987, SPE Paper 16948.
Byrnes, A.P., 1997. Reservoir characteristics of low-permeability sandstones in the Rocky Mountains. Mountain Geologist 34, 39–51.
Gonfalini, M., 2005. Formation Evaluation Challenges in Unconventional Tight Hydrocarbon Reservoirs SPE Italian Section.
Holditch, S.A., 2006. Tight gas sands. Journal of Petroleum Technology 58, 84–90.
Ingrid Campbell, Lakes Oil, 2009. An overview of Tight Gas Resources in Australia, Petroleum Exploration Society of Australia NEWS, pp. 95–100.
Law, B.E., Curtis, J.B., 2002. Introduction to unconventional petroleum systems. American Association of Petroleum Geologists Bulletin 86, 1851–1852.

Rezaee, R., Saeedi, A., Clennell, B., 2012. Tight gas sands permeability estimation from mercury injection capillary pressure and nuclear magnetic resonance data. Journal of Petroleum Science and Engineering 88–89, 92–99.
Sharifzadeh, A., 2007. Tight gas resources in Western Australia. Petroleum W.A Magazine, pp. 28–31.

CHAPTER 3

Tight gas sandstones rock typing

3.1 Introduction

As with the conventional reservoirs, tight sandstone reservoirs show some degree of heterogeneity related to geological and petrophysical properties. The reservoir heterogeneity changes from microscale within the pore system to large (field and regional) scales. One of the techniques to unravel the reservoir heterogeneity is rock typing through investigation, description, and interpretation of sedimentary and petrophysical characteristics of reservoir rocks by which different types of reservoir rocks or rock types are identified. Also, other aspects of the reservoir such as geochemical, geomechanical, seismic, and reservoir engineering parameters can be used for reservoir rock typing. Thus, any meaningful classification of reservoir rocks to differentiate and describe them based on the special characteristics of the reservoir is attributed to rock typing (Kadkhodaie-Ilkhchi and Kadkhodaie-Ilkhchi, 2018). Although rock typing methods and techniques are different based on the nature of the data and the study scale, they can be related to each other in a hierarchical framework to depict the reservoir heterogeneity, reservoir zones, and fluid flow within the reservoir (Fig. 3–1). This framework describes the application of different data sources related to static and dynamic reservoir properties in five stages as follows.

1. A direct description of reservoir rocks based on core/cutting and thin section studies, from which depositional facies are identified. Depositional facies provides information on lithology, fossil content, sedimentary textures (e.g., grain size and sorting), structures, and any feature related to depositional environment conditions. Also, the effect of post-depositional processes is reflected as diagenetic features and products within the reservoir rocks.
2. Investigation of electrofacies (log facies) by analysis of well log responses. Log data provide indirect but valuable information on the physical and compositional properties (e.g., lithology, density, and porosity) of reservoir rocks. The results from well log analysis are validated and controlled by core/cutting and thin section studies.

Reservoir Characterization of Tight Gas Sandstones
DOI: https://doi.org/10.1016/B978-0-323-90180-2.00008-6

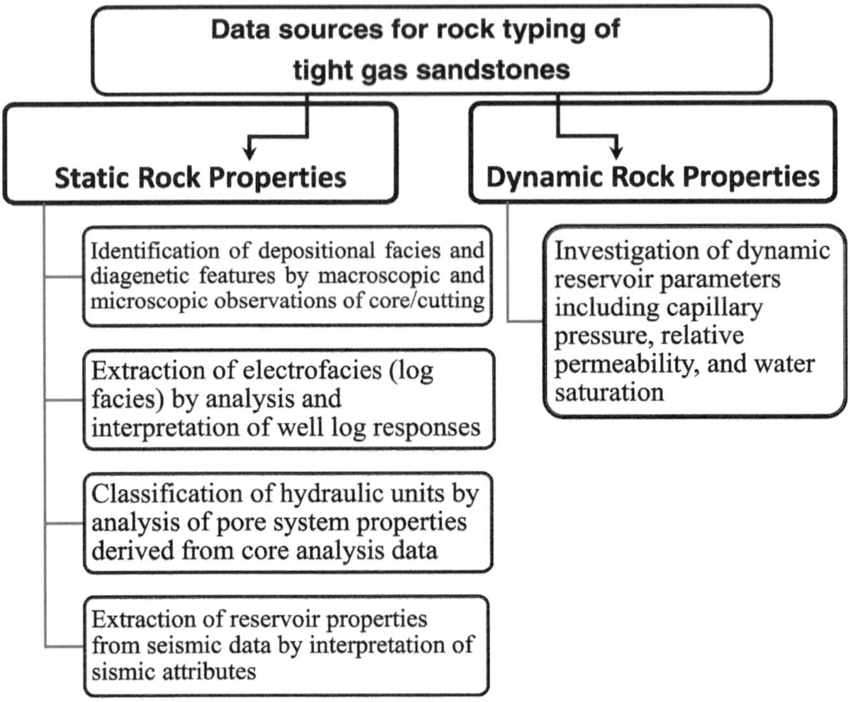

Figure 3–1 Data sources framework for reservoir rock typing of tight gas sandstones (Originally generated for this chapter).

3. Description of hydraulic units based on core porosity and permeability data. The pore system of reservoir rocks consists of pore spaces and pore throats responsible for fluid storage (porosity) and fluid flow (permeability), respectively. The relationship between these parameters is not simple, especially in heterogeneous reservoirs, and it mainly depends on many factors such as pore types, pore size, pore geometry, and the properties of fluids occupying the pore volume. Hydraulic units differentiate the rock types by finding a specific porosity-permeability relationship for each rock type.
4. Interpretation of seismic data and integration of the results with rock typing results from the core and well log data. Compared to well logs, seismic data on a larger scale but with a lower resolution, provide indirect and informative information on reservoir rocks' properties. Variation in density and porosity of geologic units is recorded on seismic data, by which reservoir heterogeneity, reservoir zones, and their distribution in the basin could be depicted. Seismic data uses seismic

attributes [e.g., acoustic impedance (AI)] to depict and classify reservoir rocks, based on which seismic facies or seismic rock types are differentiated. A set of seismic attributes can be formulated through statistical and machine learning techniques for the interpretation of reservoir rock properties such as porosity, pore type, fluids saturation, clay volume, and elastic rock properties (Kadkhodaie and Kadkhodaie, 2022).

5. Making a relationship between static rock types (depositional facies, electrofacies, seismic facies, and hydraulic units) and dynamic properties of reservoir rocks. Dynamic properties are related to the behavior of fluids in the pore system of the reservoir, and they are described based on some parameters such as capillary pressure, relative permeability, and water saturation. The results of static rock typing are reliable when they show good consistency with dynamic reservoir parameters. Therefore, each rock type is characterized by specific capillary pressure and relative permeability curve. In a practical view, considering this issue in specifying the number of rock types is significant. This means that it is not logical to estimate the number of reservoir rock types (RRTs) based solely on depositional facies when two different facies (with different textures and fossil content) have similar reservoir properties. In fact, depositional facies are a reflection of depositional environment conditions that have fundamental control on their sedimentary texture, and not necessarily on reservoir properties. Also, diagenesis may significantly modify the primary reservoir properties of depositional facies. Generally, the number of RRTs is a function of reservoir heterogeneity and complexity in pore system properties. The introduction of a limited number of rock types, mainly derived from large-scale data sources such as outcrop and seismic, does not reveal the real heterogeneity and complexity of the reservoir. On the other hand, extraction of too many rock types, mostly derived from small-scale data sources such as well logs and core/cutting and thin sections, complicates the results and encounters challenges in subsequent steps of reservoir studies such as static and dynamic reservoir modeling. A sensible choice of flow units may eliminate the need for upscaling of geological models for reservoir simulation targets (Bhattacharya et al., 2008).

The flowchart is shown in Fig. 3–2 is modified after Kadkhodaie-Ilkhchi and Kadkhodaie-Ilkhchi (2018) and demonstrates the different stages of reservoir rock typing, beginning with sedimentary rock types (SRTs) derived from core/cutting and thin section studies, then

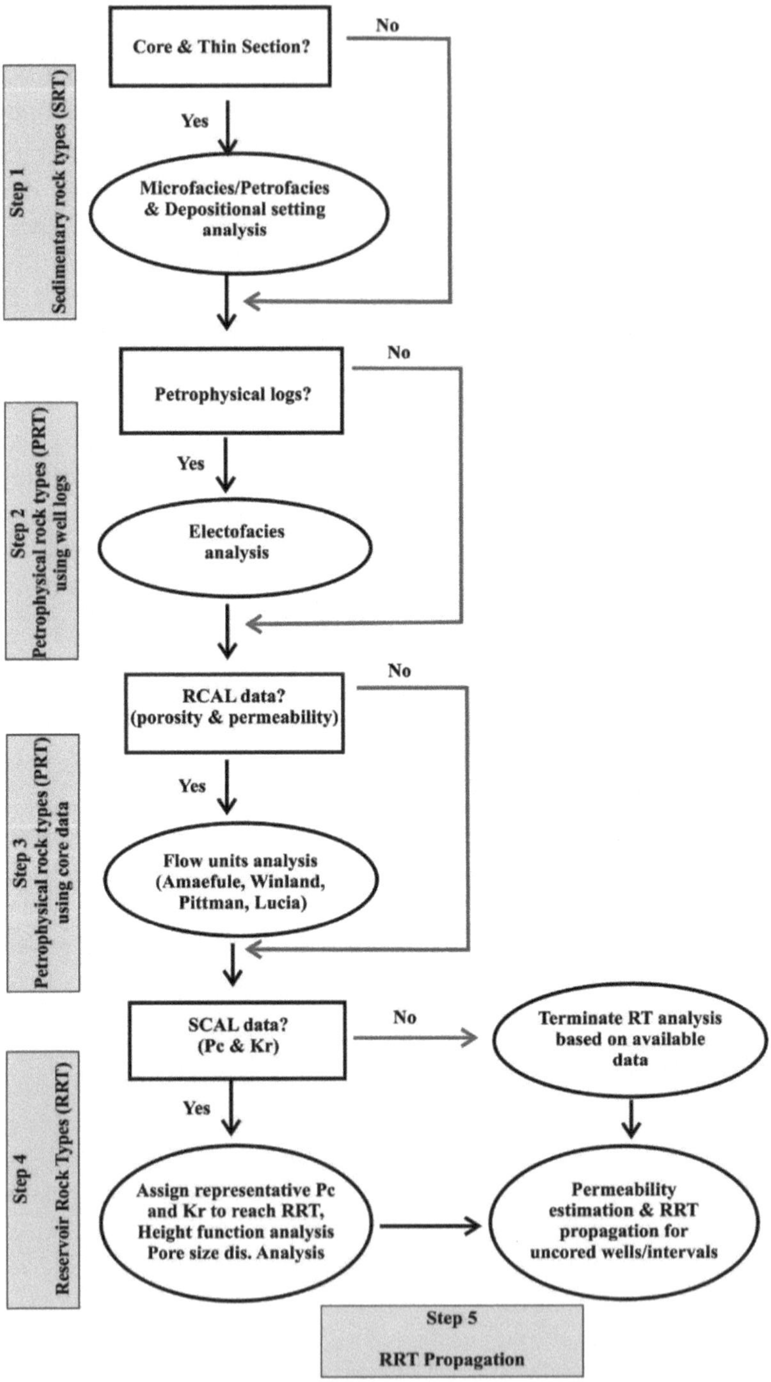

Figure 3–2 Flowchart showing different stages of reservoir rock typing. *Modified after Kadkhodaie-Ilkhchi, A., Kadkhodaie-Ilkhchi, R., 2018. A review of reservoir rock typing methods in carbonate reservoirs: relation between geological, seismic, and reservoir rock types. Iranian Journal of Oil & Gas Science and Technology 7 (4), 13–35.*

continuing with petrophysical rock types based on well logs (electrofacies) and core porosity and permeability data from RCAL [hydraulic flow units (HFUs)], and finally, it ends with RRTs from SCAL data (capillary pressure and relative permeability), and propagation of results to uncored wells and intervals.

3.2 Rock typing in tight sandstone reservoirs

In recent decades, rock typing of tight sandstone reservoirs has been the subject of many studies with the aim of establishing a relationship between the static and dynamic characteristics of the reservoir. By integrating multiple data evaluation techniques and multiple data scales, Rushing et al. (2008) introduced a core-based rock typing workflow process for low permeability (tight) sandstone reservoirs. Consequently, they introduced three main rock types, including depositional, petrographic, and hydraulic. In this workflow, depositional rock types are described based on all original large-scale sedimentary features and attitudes of reservoir rocks such as composition, texture, and structure at the time of deposition. Similarly, petrographic rock types consider the current sedimentary features (composition and texture) and also diagenesis of the rock based on pore-scale microscopic imaging. Also, hydraulic rock types that are defined based on the physical rock flow and storage properties at the pore scale, are a reflection of the current pore structure (i.e., size, geometry, and distribution of pore throats) controlling the porosity and permeability of the rock. Moreover, Aliyev et al. (2016) used different petrophysical properties such as porosity, permeability, surface area, and pore size distribution for reservoir rock typing of tight gas sandstones of Lance and Mesaverde formations from Jonah field of the Green River Basin. As a result, they identified four distinct rock types, including reservoir sandstone, non-reservoir sandstone, siltstone, and mudstone. These rock types were differentiated by examining and comparing different rock typing techniques such as flow zone indicator (FZI), effective specific surface (ESS), and hydraulic rock typing (HRT), as well as using various experimental data from analysis of mercury injection capillary pressure (MICP), nuclear magnetic resonance (NMR) and nitrogen adsorption data and acoustic velocity evaluation.

A proper rock classification of tight gas sandstones requires an understanding of depositional and diagenetic characteristics controlling the physical properties of these reservoirs (Aliyev et al., 2016). Therefore, the investigation of depositional facies and diagenetic features is the first and important step for basic reservoir studies and characterizing the reservoir heterogeneity in tight sandstone reservoirs. These attitudes and different rock typing methods are described below.

3.2.1 Depositional facies

Depositional facies that are a reflection of the conditions prevailing in the sedimentary environment at the time of deposition are characterized by various sedimentary attitudes, among which the composition (mineralogy), texture (grain size and sorting), and sedimentary structure are of special importance. In fact, lateral and vertical changes in lithofacies are derived from the spatial and temporal variations of the depositional environment (Bhaskar, 2006). The study of facies includes all macroscopic information derived from surface (outcrop) and subsurface (core/cutting) observations, as well as microscopic evidence obtained from thin section studies, SEM, and XRD analysis. Therefore, any interpretation of geological, petrophysical, and seismic characteristics of reservoir rocks requires an analysis of their depositional facies. On the other hand, pore system properties controlling the fluid flow and storage in sandstones have an intimate relationship with the facies of different depositional environments. Depositional facies control the reservoir heterogeneity of tight sandstones at both the small scale related to pore system properties and the large scale controlling the distribution of reservoir zones and their continuity in the field and inter-field areas.

Although the initial sedimentary features of tight sandstones have significantly been modified by diagenesis, some reservoir properties in these sandstones are inherited from depositional facies. For example, in low permeability sandstone reservoirs of the United States that have experienced a complex burial and uplift history, depositional facies with higher values of initial porosity and permeability are still characterized by high reservoir quality (Shanley et al., 2004). Also, by investigation of tight gas sandstones of the Upper Cambrian to Lower Ordovician Barik Member in the Khazzan region of Oman, Millson et al. (2008) indicated that, despite a prolonged and complex burial history, reservoir quality shows relationships with primary depositional texture (grain size, sorting, and clay abundance). Orsini and Rezaee's (2012) investigation of tight sandstones of the Whicher Range field of the Perth Basin in Western Australia showed that high energy and coarse-grained facies of fluvial channel and crevasse splay, despite the influence of diagenesis, have higher porosity and permeability than low energy and fine-grained argillaceous facies of channel margin and flood plain (Fig. 3–3). Furthermore, in another study in this field, Kadkhodaie et al. (2021) indicated that initial sedimentary characteristics have fundamental control over large-scale reservoir heterogeneity at the field scale, and diagenetic processes intensify this heterogeneity at the pore scale. All of these studies illustrate that depositional facies are the basic elements in the characterization of tight sandstone reservoirs, and thus having knowledge of facies and their environment is essential, because (1) on a large scale, vertical and lateral changes in depositional conditions control the thickness and continuity of reservoir rocks. Also, the extent of the reservoir container (gas-in-place) is controlled by the distribution of depositional rock types (Rushing et al., 2008). (2) any petrophysical and seismic interpretation of reservoir rocks need to be equipped with stratigraphy and sedimentology elements to depict the internal nature of the reservoir rocks, and (3) reconstruction of a depositional framework of tight

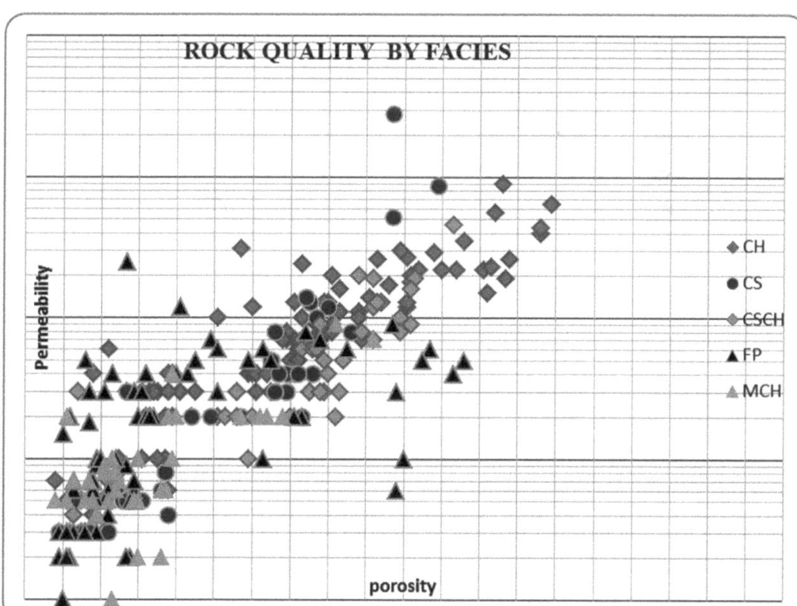

Figure 3–3 Porosity-permeability cross plot for different facies of tight sandstones in the Whicher Range field of Western Australia (Orsini and Rezaee, 2012). *CH*, channel; *CS*, crevasse splay; *CSCH*, crevasse channel; *FP*, flood plain; *MCH*, channel margins.

sandstone reservoirs helps to consider a suitable and effective strategy for development, production and simulation targets of the field.

3.2.2 Diagenetic facies

Predicting rock properties based on depositional environment and facies is reliable when the effect of diagenesis on pore system properties is not significant. But in most tight gas sands that have effectively been subjected to diagenetic processes, this prediction is challengeable. The pore system in these reservoirs has significantly been modified by diagenetic processes in terms of pore type and pore structure. The main diagenetic features affecting the pore systems in tight sandstones are compaction, cementation (type and degree), and dissolution. Clogging pores and reducing the pore throat's size by diagenetic cements such as quartz, calcite, and clay minerals can disconnect pores, increase tortuosity and restrict fluid flow (DuBois et al., 2004; Aliyev et al., 2016). Moreover, alteration and dissolution of metastable grains change primary intergranular pore spaces to secondary pore types (micropores and isolated vugs), by which the heterogeneity of reservoir rocks is increased at the pore scale. Diagenetic facies in tight sandstones can be defined as individual rock types or they can be considered as features in the context of SRTs. Also, they are used in the interpretation of petrophysical rock types such as flow units and electrofacies. The basic elements used in the definition of diagenetic facies of reservoir rocks are diagenetic features and pore types. In most cases, the textural and mineralogical maturity of depositional facies has the main control on the distribution and development of diagenetic products in the reservoir. The effect of compaction in sandstones rich in soft and ductile components (e.g., mica, shale, and volcanic rock fragments) is noticeable. Also, the presence of silica as overgrowth cement in quartz-rich sandstones and the effect of dissolution as secondary pores or micropores in feldspathic-rich sandstones are common diagenetic features (see Section 5.2 of Chapter 5). The combined effect of these processes results in a compacted and cemented framework with isolated pore types in tight sandstone reservoirs. The study of diagenetic facies in tight sandstones is useful for the description of the diagenetic history, evaluation of reservoir quality, and prediction of sweet spots (Wang et al., 2019). Lai et al. (2018) provided a review of diagenetic facies in tight sandstones and established a model for the prediction of diagenetic facies from well logs. They considered similar compositional and textural attributes, matrix, cement, and porosity systems to define a given diagenetic facies. Similarly, Li et al. (2021) developed a KNN algorithm for the automatic identification of diagenetic facies from well logs in tight sandstone reservoirs of the Lower Silurian Kepingtage Formation in the Tarim Basin of China. They considered two parameters of cause (type and intensity) and effect (influence on the reservoir) of diagenetic processes, based on which three tight diagenetic facies related to the dominant effect of compaction, carbonate cementation, and bitumen filling, and three porous diagenetic facies mostly related to the effect of dissolution were recognized.

3.2.3 Tight gas sandstone electrofacies

The concept of electrofacies first was introduced by Serra and Abbott (1980) to characterize a sedimentary unit by well log responses. A set of well logs provides indirect but valuable information on different aspects of a reservoir, including mineralogy, sedimentary texture and structure, and petrophysical properties such as porosity, permeability, and fluid content (Lee and Dutta-Gupta, 1999). Although electrofacies are not necessarily a reflection of specific lithology or depositional environment of reservoir rocks, the produced classes are an indication of reservoir properties. Unlike other rock typing methods that rely on cuttings, cores, and outcrops information, electrofacies are described based on well log responses (Davis, 2018). This is an important aspect of electrofacies, which deals with the continuous data that is available in most wells of a field and is also correlatable between them. It should be noted here, that each electrofacies that is a reflection of specific physical and compositional characteristics of a rock interval, should finally be controlled by information obtained directly from core/cutting observations. The first step in electrofacies analysis of a field with many wells is to select the key wells with the appropriate log data and, if possible, core data and information. In these wells, finding a common set of logs available in all wells, and covering the interesting interval, is the next step. Commonly used logs that provide information on lithological and petrophysical characteristics of reservoir rocks include gamma-ray (GR, API), neutron (NPHI, v/v), sonic (DT, μs/ft), and density (RHOB, g/cm^3). Then, these logs should be checked and controlled for some corrections such as cable tension effects for adaptation of core and well log data, removal of bad hole flags (BHF) data, null values, and log tails. Also, due to differences in scale and unit of different logs, they can be unified and standardized. Log standardization is accomplished by subtracting the mean log response from every log reading and dividing the remainder by the standard deviation of the response in the interval of interest (Davis, 2018). There are different methods and techniques for the investigation of electrofacies. These methods are identical for clastics and carbonates and also for conventional and unconventional reservoirs. Clustering of well log data by statistical and mathematical methods is one of the commonly used techniques that provide different classes of well log responses or electrofacies. The number of electrofacies depends basically on the degree of heterogeneity of the reservoir rocks. The optimal number can usually be inspected by a trial and error method. In general, when a relatively good consistency between electrical logs and reservoir rock properties was attained, the results is acceptable. Extraction of many electrofacies from input log data is a common phenomenon in the resultant output data derived from the software used. In this case, optimizing the number of electrofacies is accomplished by merging the electrofacies with average values of well log responses. In tight reservoir sandstones, under the effect of diagenesis and in some cases heterolithic lithology, making an exact correlation between the identified electrofacies and depositional facies may be impossible. This is not a serious problem in the characterization of reservoir rocks as electrofacies generally identify rock types based on their reservoir properties, independent of their depositional texture. Accordingly, the judgment of reservoir characteristics solely based on sedimentary parameters of tight reservoirs may be misleading. A workflow for extraction and study of electrofacies from well log data is shown in Fig. 3–4. Correia and Schiozer (2016) used electrofacies analysis in the study of fine-scale reservoir heterogeneity and delineating stratigraphic barriers of the Early–Middle Jurassic sandstone reservoir of the Norne Field (offshore Norway). Moreover, Kadkhodaie-Ilkhchi et al. (2013) utilized the electrofacies concept in association with HFUs for

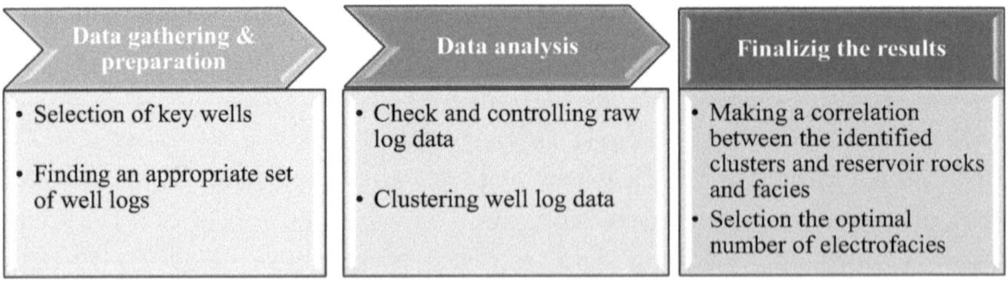

Figure 3–4 Workflow showing the different stages of extracting electrofacies from well log data based on cluster analysis.

reservoir characterization of tight gas sandstones of the Whicher Range field of Western Australia. As a result, they introduced three tight rock types that are distinguished based on the intensity of diagenetic effects and the values of porosity and permeability. Also, the application and importance of electrofacies in the characterization of other unconventional reservoirs such as shale gas have been identified (Karimian-Torghabeh et al., 2014; Mukherjee et al., 2017; Stinco and Barredo, 2021).

3.2.4 Hydraulic flow units

Evaluation of factors affecting fluid flow in reservoir rocks is accomplished by different techniques, among which HFUs are of significant importance. This concept considers the heterogeneity of reservoir rocks by characterizing their pore system based on the core porosity and permeability relationships. Therefore, the reservoir interval can be divided into different units or zones that are correlatable and mappable, and each unit shows similar averages of petrophysical rock properties controlling fluid flow (Amaefule et al., 1993). Furthermore, the subdivision of the reservoir rock volume into units with similar properties for fluid flow is appropriate for flow simulation studies (Bhattacharya et al., 2008). This concept in reservoir characterization has widely been used by many authors in conventional (e.g., Amaefule et al., 1993; Aggoun et al., 2006; Kadkhodaie-Ilkhchi and Amini, 2009; Riazi, 2018; Yu, 2021) and unconventional reservoirs such as tight sandstones (e.g., Clarkson et al., 2012; Aguilera, 2014; Opuwari et al., 2021). In conventional reservoirs, the most common method used to determine flow units is the FZI concept introduced by Amaefule et al. (1993), which can be derived from Eq. (3.1).

$$\mathrm{FZI} = \frac{\mathrm{RQI}}{\mathrm{PMR}} \tag{3.1}$$

where RQI is reservoir quality index (μm) and PMR is pore to matrix ratio that is defined by Eqs. (3.2) and (3.3), respectively.

$$\mathrm{RQI} = 0.0314\sqrt{\frac{K}{\varphi}} \tag{3.2}$$

$$\mathrm{PMR} = \varphi/(1-\varphi) \tag{3.3}$$

In these equations, k is permeability, mD; and φ is fractional porosity.

Each flow unit is characterized by a narrow range of FZI values, and it shows a unique relationship on the porosity-permeability plot. Primary depositional facies and diagenesis have significant control on the pore system properties of the identified flow units. In general, flow units with higher values of RQI and FZI (high-grade or high-quality flow units), are related to facies with larger pore sizes and good connectivity of pore network, and so they have a high potential for fluid flow. In contrast, facies with small pore sizes and poor connectivity of pore networks are related to low-grade (low quality) flow units. Kadkhodaie et al. (2021) considered the FZI method in the characterization of tight sandstones in the Whicher Range field of Perth Basin in Western Australia. The core porosity and permeability plot for the identified flow units is shown in Fig. 3–5. They found a good consistency between the identified flow units and their related depositional facies. According to their study results, low-grade flow units (HFU-1 & HFU-2) show a wide range of porosity and very low permeability covering fine and very fine-grained argillaceous sandstones and siltstones with a pore system mainly consisting of micropores and poorly connected pores. In contrast, high-grade flow units (HFU-3 &

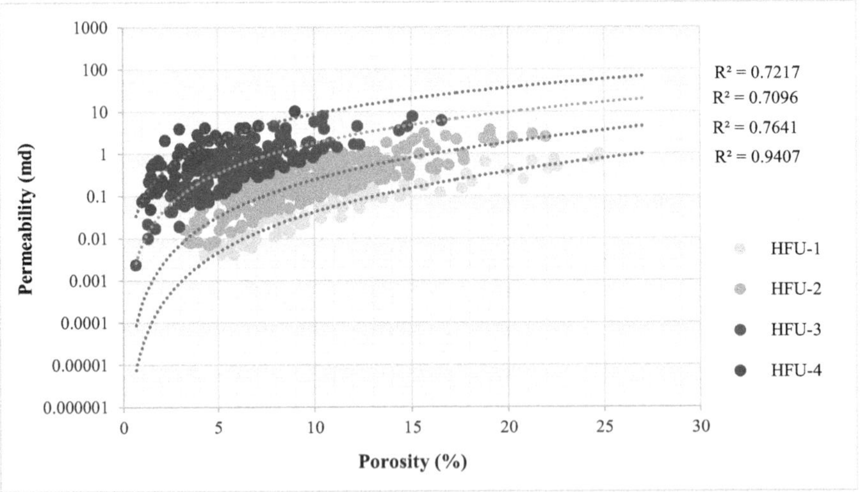

Figure 3–5 Core porosity and permeability plot for the identified flow units in tight sandstone of the Whicher Range field of Perth Basin, Western Australia. *Modified after Kadkhodaie-Ilkhchi, R., Rezaee, R., Moussavi-Harami, R., Kadkhodaie-Ilkhchi, A., 2013. Analysis of the reservoir electrofacies in the framework of hydraulic flow units in the Whicher Range field, Perth Basin, western Australia, Journal of Petroleum Science and Engineering, 111, 106–120.*

HFU-4) with a narrow range and porosity and higher permeability are related to coarse and very coarse-grained sandstones consisting of pores with a relatively better connection.

3.3 Seismic-based rock typing

Seismic data are an effective tool in characterizing reservoir rocks, by which a profound understanding of large-scale reservoir heterogeneity and distribution of reservoir zones in the field is obtained. Estimation of petrophysical properties of reservoir rocks such as porosity, water saturation, and pore-throat size by correlating seismic attributes and well log data that have also been calibrated with core data is of particular importance in the oil industry so that many researchers have considered it in evaluating reservoir properties. A comprehensive discussion of the application of seismic attributes in the delineation of reservoir properties and detection of gas zones in carbonate and sandstone reservoirs has been provided by Kadkhodaie and Kadkhodaie (2022). The application of seismic data in reservoir characterization of tight sandstone reservoirs has been considered by many researchers. As Li et al. (2004) used seismic attributes in the identification of gas-bearing horizons of a tight sandstone reservoir in the Majing gas field of Sichuan Basin in China. Moreover, Cheng et al. (2012) estimated pore microstructure properties from seismic data by establishing 3D rock-physics templates (RPTs) in tight-gas sandstone samples of the Xujiahe Formation in Western Sichuan Basin of China. Furthermore, some researchers by making a relationship between seismic elastic parameters and core-derived diagenetic features tried to investigate the spatial distribution of diagenetic facies in tight sandstones (e.g., Zhang et al., 2018; Wang et al., 2022). Kadkhodaie-Ilkhchi et al. (2014) used AI derived from seismic inversion for porosity evaluation of tight sandstones of the Whicher Range field of the Perth Basin in Western Australia. They concluded that variations in AI could be well correlated with depositional and diagenetic characteristics of tight sandstones. As a result, by analysis of AI, three sandstone rock types with high, medium, and low AI values were identified. A seismic section of AI for tight sandstones of the field is shown in Fig. 3–6. According to this figure, a clear trend in the distribution of porous and low- to non-porous zones in the field is observed. High AI ($AI > 38000$) sandstones with average porosity of 5% are concentrated at the base of the reservoir interval and are gradually changed to medium AI sandstones ($27000 < AI < 38000$) with average porosity of 7% in the middle interval followed by low AI

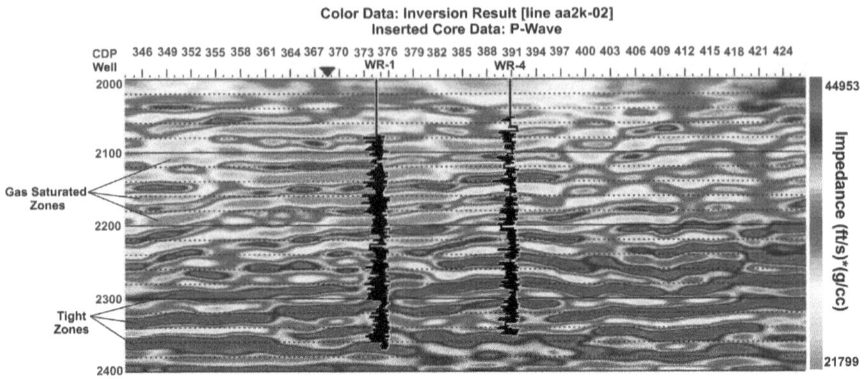

Figure 3–6 A seismic section of acoustic impedance derived from seismic inversion for tight gas sandstones of the Whicher Range Field in the Perth Basin of Western Australia.

sandstones (AI < 27000) with average porosity of 10% at the top of the reservoir. In further efforts to the identification of RRTs from seismic data in the Whicher Range field, AI in combination with velocity deviation log was used for pore system analysis and identification of gas-saturated zones in the field (Kadkhodaie et al., 2021; Kadkhodaie and Kadkhodaie, 2022).

3.4 Fluid flow and hydraulic rock types in tight gas sandstones

A comprehensive understanding of tight sandstone rock types compatible with production data, and practical for field development, requires an analysis of geological and physical rock properties under reservoir conditions. According to the results of different studies (e.g., Rushing et al., 2008; Aliyev et al., 2016; Ji et al., 2022), the most important parameter in the delineation of tight sands rock types is the pore-throat diameter that controls the fluid flow. Therefore, the identification of hydraulic rock types (HRT) that consider the dominant pore-throat diameter, shows a good consistency of permeability-porosity correlations for each rock type. Also, any description of hydraulic rock types should consider the in-situ reservoir conditions. This is because, in tight sandstone reservoirs, pore systems include a narrow range of pore sizes with small pore-throat diameters that make them very sensitive to confining pressure and water saturation.

Many different factors control the fluid flow in tight sandstone reservoirs. In fact, fluid flow in tight sandstone reservoirs is intimately related to depositional elements, pore networks, and petrophysical properties by which hydraulic rock types are described (Fig. 3–7). These factors are also closely related to each other and affect the flow rate and production of a field.

Depositional elements that account for internal reservoir heterogeneity related to facies and lithology changes, affect fluid flow in the reservoir (Pranter et al., 2007). Lateral and vertical variations in depositional facies and structures can limit the distribution and continuity of sedimentary units, which in turn affect the distribution of reservoir zones, and increase the reservoir heterogeneity. Permeability is mainly controlled by the pore-throat size rather than the total porosity. This is because porosity is a function of pores volume, and is independent of their structure. However, permeability is related to pore geometry (size and shape), distribution of pores, and their connectivity and surface area of the grains in contact with the flowing fluid (Aliyev et al., 2016). Petrophysical properties such as water saturation have the main effect on relative permeability and hydrocarbon saturation. Also, water saturation facilitates the development of diagenetic processes such as cementation within the reservoir. For example, in tight sandstones of the Late Cretaceous Almond Formation in Wamsutter field of Wyoming, Tobin et al. (2010) indicated that despite the presence of hydrocarbons in the pore system reducing the rate of cementation, destructive diagenesis continued to decrease the pore system in the presence of enough water saturation.

Clarkson et al. (2012) in their studies on a tight gas reservoir of the Lower Triassic Montney Formation in the Peace River Basin, addressed the deficiencies of routine core analysis in terms of reservoir heterogeneity as well as in-situ reservoir conditions. They introduced a non-routine method by integrating the results from profile permeability measurements corrected to in-situ stress conditions and dominant pore size controlling flow

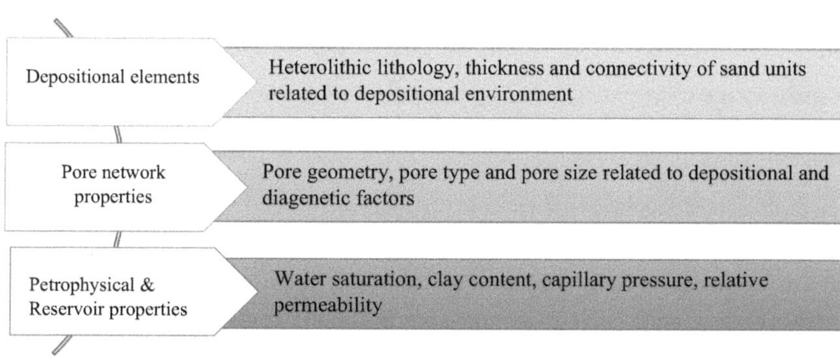

Figure 3–7 Factors controlling the fluid flow in tight sandstone reservoirs.

derived from low-pressure N2 adsorption isotherms. They mentioned that the profile permeability, for which data are collected at a finer resolution than the routine core data, is an effective method for characterizing small-scale reservoir heterogeneity, but in this method also there is still the problem related to in-situ reservoir conditions. For this purpose, they used pulse-decay permeability measured at reservoir overburden stress conditions. A part of this dependence of permeability on confining pressure was attributed to the presence of micro-fractures or slot pores in the samples.

Therefore, the heterogeneous nature and complexity of pore structure of tight gas sandstone reservoirs have encountered reservoir rock typing with challenges in these reservoirs. Also, in most cases, considering a certain method for a specific tight sandstone reservoir does not mean it is applicable and suitable for unraveling the reservoir heterogeneity in other tight sandstone reservoirs. A look at the literature shows that in recent years, several researchers have tried to provide effective and efficient methods for rock typing of these reservoirs. Although the proposed methods are limited to specific reservoirs, the results illustrate the challenges associated with the complexity of these reservoirs and provide the framework for future researchers to address these complexities in their works. In most cases, a combination of different methods and techniques considering the lithological, petrophysical, and reservoir characteristics is proposed. Aliyev et al. (2016) investigated hydraulic units using three rock typing techniques, including FZI, ESS, and HRT. According to their study results, FZI correlates permeability with porosity based on pore geometry factors (i.e., pore shape factor (F_s), tortuosity (τ), and specific surface area per unit grain volume (Sv_g)), has uncertainty in low porosity and permeability rocks. Also, the ESS method correlates permeability and porosity with a specific surface area of grains exposed to fluid flow (S_g). By considering the pore space as orthogonal tubes, Mortensen et al. (1998) described the term active porosity, Eq. (3.4), as the main part of the pore space that participates in fluid flow.

$$C(\varphi) = \left(4\cos\left\{\frac{1}{3}\arccos(2\varphi - 1) + \frac{4}{3}\pi\right\} + 4\right)^{-1} \tag{3.4}$$

Then, by replacing tortuosity (τ) and shape factor (F_s) in Cozeny-Karman equation with active porosity, they introduced Eq. (3.5). Therefore, the ESS (S_{g-eff}) as an independent parameter can be used for differentiation of rock types on the porosity-permeability plot. Specific surface is determined by BET nitrogen adsorption method.

$$K = c\,(\varphi)\frac{1}{S_{g-\text{eff}}^{2}}\frac{\varphi^{3}}{(1-\varphi)^{2}} \tag{3.5}$$

Aliyev et al. (2016) connected the magnitude of $S_{g-\text{eff}}$ to clay content. They indicated that this parameter has inverse relationship with reservoir quality, and high clay content has significant effect on increasing ESS and decreasing effective porosity. As with the FZI method, their approach shows uncertainty in low porosity and permeability rocks. In contrast, HRT makes a correlation between porosity and permeability based on the main pore-throat radius dominating the fluid flow. Dominant pore-throat radius is determined by Thomeer's (1960) method, which uses MICP data by plotting injected mercury saturation (S_{hg}) against the ratio of saturation to capillary pressure. Then, the apex point mercury saturation, in which fluid flow is continuous, is determined, and the corresponding pore-throat is specified by cumulative mercury saturation and pore throat radius plot. Despite FZI and $S_{g-\mathit{eff}}$ methods, there is no limitation and uncertainty for HRT-based method in low porosity and permeability rocks (Aliyev et al., 2016). Moreover, Lis-Śledziona (2019) by the study the Cambrian tight sandstone of the Baltic Basin considered and compared three reservoir rock typing methods, including heterogeneous rock analysis (HRA), FZI, and a new flow unit discriminator named tight reservoir quality index (TRQI). A summary of the results from this study is discussed below. The HRA method was used to cluster well log-derived data calibrated with core-measured data such as shale volume, pore size, effective porosity, and permeability. As a result, four clusters of flow units (HRA1 to HRA2) were recognized. HRA1 with the best reservoir properties (porosity, permeability, and pore size) and HRA4 with the lowest of these properties were classified as sub-tight and very tight units, respectively. HRA2 and HRA3 showed the intermediate properties of the mentioned units. Also, based on the FZI method, four flow units (FZI unit 1 to FZI unit 4) were identified. A decrease in permeability was observed from flow unit 1 toward flow unit 3. However, flow unit 4 showed very high permeability related to the presence of fractures. Also, the proposed TRQI is calculated based on Eq. (3.6).

$$\text{TRQI} = \frac{P * \sqrt{\frac{K}{PHIE}}}{SWI_{C}} \tag{3.6}$$

Where *P* is the size of pores (μm), *K* is permeability (md), *PHIE* is effective porosity (in fraction), and SWI_C is capillary water saturation (in fraction). Based on this method, four reservoir units were identified, which are in descending order of reservoir quality sub-tight (TRQI 1> 15), sub-tight to tight (TRQI 2 = 3.8–15), tight (TRQI 3 = 1–3.8), and very tight (TRQI = 0–1). Finally, Lis-Śledziona (2019) by comparing the results of the three methods concluded that HRA and TRQI are proper methods for reservoir rock typing in tight reservoirs. In contrast, FZI is not a good-quality index. This is because a certain flow unit that includes an entire range of porosity and permeability causes low porosity and low permeability rocks to be classified with high porosity and permeability rocks in one FZI group, as indicated by Aliyev et al. (2016). In these reservoirs containing a large amount of small and disconnected pores, the correlation between porosity and permeability is weak, and disconnected pores have no significant contribution to permeability and fluid flow. Another reason mentioned by Lis-Śledziona (2019) is very little change in FZI in low porosity and low permeability zones resulting in the weak separation of flow units. On the other hand, Ji et al. (2022) emphasized the non-linear and non-Darcy flow derived from numerous nanopores in tight sandstones, which cannot be treated by traditional rock typing methods such as FZI and R_{35} applicable to conventional reservoirs. Therefore, they proposed three rock typing indexes, including Darcy index (Fi), gas non-Darcy index (Kc), and liquid non-Darcy index (Sr), combined with the weighted fuzzy kNN algorithm to reach an improved rock typing method that considers different flow characteristics in the multiscale pore structure of tight sandstones. These indexes are briefly described below. They derived the Darcy index (Fi) from Eq. (3.7).

$$\mathrm{Fi} = \sqrt{\frac{\mathrm{Kd}}{\varphi}} = \frac{R_{\mathrm{nt}}}{\sqrt{8\tau}} \tag{3.7}$$

In this equation, Kd is Darcy permeability derived from Eq. (3.8), and R_{nt} is determined by Eq. (3.9).

$$\mathrm{Kd} = \frac{\varphi_f R_{\mathrm{nt}}^2}{8\tau} \tag{3.8}$$

where φ_f is the flowing porosity and τ is tortuosity.

$$R_{nt} = \left(\frac{1}{N}\sum_{i=1}^{N}\left(\frac{1}{r_i}\right)^{Df-3}\right)^{\frac{1}{Df-3}} \tag{3.9}$$

where r_i, is the ith pore radius, N is the total number of pores and D_f is the fractal dimension of the pore surface. Ji et al. (2022) indicated Fi which is a function of tortuosity and R_{nt} related to pore and pore throat properties, is more suitable than *FZI* for tight sandstone rock typing. Also, non-Darcy indexes for gas and liquid characterize nanopores responsible for non-Darcy flow in tight sandstones. The gas non-Darcy index (Kc) and the liquid non-Darcy index (Sr) are derived from Eq. (3.10) and Eq. (3.11), respectively.

$$Kc = \frac{K_{app_{gas}}}{Kd} - d \tag{3.10}$$

Where $K_{app_{gas}}$ is apparent permeability for gas, derived from Jiang et al.'s (2017) formula, accounting for slip flow and Knudsen diffusion in tight sandstones.

$$Sr = \frac{K_{app_{liquid}}}{Kd} - d \tag{3.11}$$

where $K_{app_{liquid}}$ is apparent permeability for liquid, derived from Afsharpoor and Javadpour (2016) equation. Ji et al. (2022) concluded that pores radius for gas non-Darcy flow is smaller than that of liquid non-Darcy flow, and both are smaller than the Darcy flow. As a result, they recognized three main parts related to pore sizes (R_{Kc} and R_{Sr}) controlling fluid flow in tight sandstones. When pores radius is smaller than R_{kc}, both non-Darcy flow for gas and liquid are dominant, and for pore radius greater than R_{kc} and less than R_{Sr}, only liquid non-Darcy flow is dominant. When the pore radius is greater than R_{Sr}, the fluid flow follows Darcy's law.

In a summary of the results of different rock typing methods in tight sandstone reservoirs, it can be stated that due to the complex pore structure and strong heterogeneous nature of these reservoirs, common rock typing methods such as FZI are used in conventional sandstones do not provide comprehensive results. This is attributed to some factors such as non-Darcy flows due to the dominant presence of nano- to micro-pores and also disconnected pores that have no significant contribution to fluid flow resulting in the poor correlation of porosity and permeability in tight reservoirs. Also, due to the effect of diagenesis on the pore system, and the narrow range of permeability, discrimination between different flow units based on porosity and permeability is not complete. Therefore, in tight reservoirs focusing on rock typing methods based on the pore throat sizes (i.e. HRT), which considers the correlation between pore throat size and permeability, provides a more realistic picture of microscale reservoir heterogeneity and RRTs which is compatible with the dynamic behavior of the reservoir. On the other hand, relying on data that are in agreement with the in-situ conditions of the reservoir, such as pressure and water saturation, is of considerable importance in the reliability of rock typing results and fluid flow simulation in these reservoirs. Moreover, the use of rock typing methods in the study of heterogeneity of pore system and reservoir characteristics should be in the framework of heterogeneity related to the lithological properties (texture and composition) and diagenesis of reservoir sandstones to respond to current and future production challenges of the field. It should be noted that although tight sandstones are generally similar in reservoir properties, there are various types of tight sandstones with regard to basinal setting and diagenetic history. Therefore, the selection of a specific rock typing method for a certain tight sandstone reservoir may not be suitable for others.

3.5 Classification of tight gas sandstones

3.5.1 Depositional/diagenetic classification

Based on the depositional, compositional, and diagenetic features, we classify tight sandstones into two main types including sedimentary (Type S) and diagenetic (Type D) types (Fig. 3–8). Sedimentary/depositional type tight sandstones are rich in graywackes and their depositional texture has a significant control on reservoir properties. In contrast, diagenesis plays a key role in pore system properties in diagenetic type tight sandstones. Based on the mineralogy and predominant diagenetic features, diagenetic types are further divided into three groups, including Type D_Q (quartz dominant), Type D_F (feldspar dominant), and Type D_L (lithic dominant). Type D_Q includes quartz arenite sandstones with quartz overgrowth cement as the dominant diagenetic feature. Clay cements with minor importance and compaction are also the other diagenetic features in this type. Type D_F includes feldspathic sandstones, in which the presence of feldspars plays an important role in the development of diagenetic processes such as authigenic clays, carbonate cements, and dissolution. Type

Depositional/diagenetic classification of tight sandstones				
Sandstone Type	Diagenetic Type			Sedimentary Type
	Type D_Q	Type D_F	Type D_L	Type S
Petrofacies	Quartz arenite	Feldspathic arenite	Lithic arenite	Greywacke
Textural Maturity	Mature to Sub-mature	Immature to Sub-mature	Immature to Sub-mature	Immature
Dominant Diagenetic feature	Cementation (quartz overgrowth, clay cement) Compaction	Cementation (clay, calcite, feldspar overgrowth) Alteration Dissolution Compaction	Compaction Alteration Dissolution	Compaction
Example	Cambrian tight Sandstone of Baltic Basin	Late Permian tight sandstone of Perth Basin, Australia	Lower Silurian tight sandstone of Tarim Basin, China	Eocene-Miocene tight sandstone of Moghan Basin, Iran

Figure 3–8 Different types of tight sandstone reservoirs based on depositional facies, mineralogy, and dominant diagenetic features.

D_L consists of litharnite sandstones in which the effect of diagenesis is observed in the form of extreme compaction, alteration, and dissolution.

3.5.2 Reservoir quality-based classification

Based on the reservoir permeability, we classify tight sandstone reservoirs into four main groups sub-tight, tight, very tight, and ultra-tight (Fig. 3–9). Each group is further classified into two classes based on the reservoir porosity (Class I to VIII).

Sub-tight sandstones have a permeability of more than 0.1 mD and less than 1 mD. Based on the porosity, they are classified into Class I (porosity between 0.05 and 0.1) and Class II (porosity < 0.05).

Tight sandstones have a permeability of more than 0.01 mD and less than 0.1 mD. Based on the porosity, they are classified into Class III (porosity between 0.05 and 0.1) and Class IV (porosity < 0.05).

Very Tight sandstones have a permeability of more than 0.001 mD and less than 0.01 mD. Based on the porosity, they are classified into Class V (porosity between 0.05 and 0.1) and Class VI (porosity < 0.05).

Ultra-Tight sandstones have a permeability of less than 0.001 mD. Based on the porosity, they are classified into Class VII (porosity between 0.05 and 0.1) and Class VIII (porosity < 0.05).

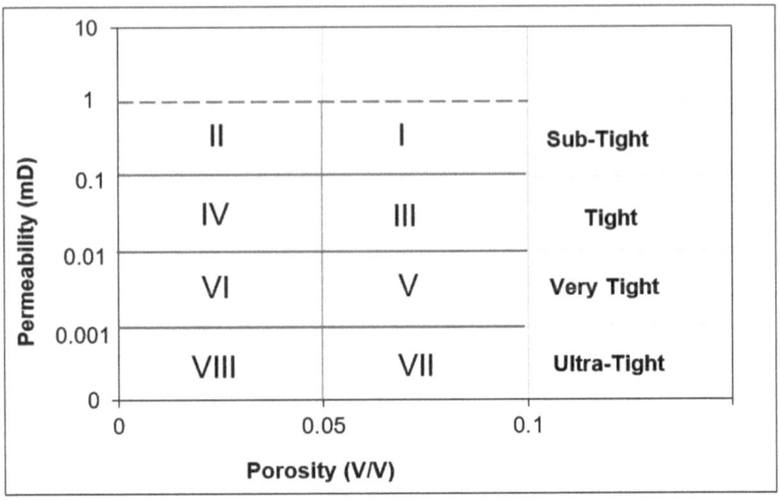

Figure 3–9 Classification of tight sandstones based on porosity and permeability.

It should be noted that due to the heterogeneous nature of tight sandstones, a tight gas reservoir can include multiple groups and classes of the introduced classification.

References

Afsharpoor, A., Javadpour, F., 2016. Liquid slip flow in a network of shale noncircular nanopores. Fuel 180, 580–590.

Aggoun, R.C., Tiab, T., Owayed, J.F., 2006. Characterization of flow units in shaly sand reservoirs Hassi R'Mel Oil Rim, Algeria. Journal of Petroleum Science and Engineering 50, 211–226.

Aguilera, R., 2014. Flow units: from conventional to tight-gas to shale-gas to tight-oil to shale-oil reservoirs. SPE Reservoir Evaluation Engineering 17, 190–208.

Aliyev, E., Saidian, M., Prasad, M., Russell, B., 2016. Rock typing of tight gas sands: a case study in lance and mesaverde formations from jonah field. Journal of Natural Gas Science and Engineering 33, 1260–1270.

Amaefule, J.O., Altunbay, M., Tiab, D., Kersey, D.G., Keelan, D.K., 1993. Enhanced reservoir description: using core and log data to identify hydraulic (flow) units and predict permeability in uncored intervals/wells. In: Proceedings of the SPE Paper 26436, p. 1–16.

Bhaskar, G.B., 2006. Electro lithofacies analysis for depositional history and stratigraphy of Manuguru Coalfield using geophysical well logs. Journal of Indian Geophysical Union. 10 (3), 241–254.

Bhattacharya, S., Byrnes, A.P., Watney, W.L., Doveton, J.H., 2008. Flow unit modeling and fine-scale predicted permeability validation in atokan sandstones: Norcan east field, Kansas. AAPG Bulletin 92, 709–732.

Cheng, W., Ba, J., Carcione, J.M., Pang, M., Wu, C., 2012. Estimation of the pore microstructure of tight-gas sandstone reservoirs with seismic data. Frontiers in Earth Science 9, 646372.

Clarkson, C.R., Jensen, J.L., Pedersen, P.K., Freeman, M., 2012. Innovative methods for flow-unit and pore-structure analyses in a tight siltstone and shale gas reservoir. AAPG Bulletin 96 (2), 355–374.

Correia, G.G., Schiozer, D.J., 2016. Reservoir characterization using electrofacies analysis in the sandstone reservoir of the norne field (offshore Norway). Petroleum Geoscience 22 (2), 165–176.

Davis, J.C., 2018. Electrofacies in reservoir characterization. In: Daya Sagar, B., Cheng, Q., Agterberg, F. (Eds.), Handbook of Mathematical Geosciences. Springer, Cham. Available from: https://doi.org/10.1007/978-3-319-78999-611.

DuBois, D., Wynne, P., Smagala, T., Johnson, J., Engler, K., McBride, B., 2004. Geology of Jonah Field, Sublette County, Wyoming.

Ji, L., Lin, M., Jiang, W., Cao, G., Xu, Z., Hao, F., 2022. An improved rock typing method for tight sandstone based on new rock typing indexes and the weighted fuzzy kNN algorithm. Journal of Petroleum Science and Engineering 210, 109956.

Jiang, C.B., Duan, M.K., Yin, G.Z., Wang, J.G., Lu, T.Y., Xu, J., Zhang, D.M., Huang, G., 2017. Experimental study on seepage properties, AE characteristics and energy dissipation of coal under tiered cyclic loading. Engineering Geology 221, 114–123.

Kadkhodaie, A., Kadkhodaie, R., 2022. Acoustic, density, and seismic attribute analysis to aid gas detection and delineation of reservoir properties. In: Wood, D.A., Cai, J. (Eds.), Sustainable Geoscience for Natural Gas Subsurface Systems, 2. p. 434.

Kadkhodaie, R., Kadkhodaie, A., Rezaee, R., 2021. Study of pore system properties of tight gas sandstones based on analysis of the seismically derived velocity deviation log: a case study from the Perth Basin of western Australia. Journal of Petroleum Science and Engineering 196, 108077.

Kadkhodaie-Ilkhchi, A., Amini, A., 2009. A fuzzy logic approach to estimating hydraulic flow units from well log data: a case study from the Ahwaz oil field, south Iran. Journal of Petroleum Geology 32, 67–78.

Kadkhodaie-Ilkhchi, A., Kadkhodaie-Ilkhchi, R., 2018. A review of reservoir rock typing methods in carbonate reservoirs: relation between geological, seismic, and reservoir rock types. Iranian Journal of Oil & Gas Science and Technology 7 (4), 13–35.

Kadkhodaie-Ilkhchi, R., Moussavi-Harami, R., Rezaee, R., Nabi-Bidhendi, M., Kadkhodaie-Ilkhchi, A., 2014. Seismic inversion and attributes analysis for reservoir quality evaluation of the tight gas sandstones of the Whicher Range field in the Perth Basin, Western Australia. Journal of Natural Gas Science and Engineering 21, 1073–1083.

Kadkhodaie-Ilkhchi, R., Rezaee, R., Moussavi-Harami, R., Kadkhodaie-Ilkhchi, A., 2013. Analysis of the reservoir electrofacies in the framework of hydraulic flow units in the Whicher Range field, Perth Basin, western Australia. Journal of Petroleum Science and Engineering 111, 106–120.

Karimian-Torghabeh, A., Rezaee, R., Moussavi-Harami, R., Pradhan, B., Kamali, M.R., Kadkhodaie, A., 2014. Electrofacies in gas shale from well log data via cluster analysis: a case study of the Perth Basin, Western Australia. Central European Journal of Geosciences 6 (3), 393–402.

Lai, J., Wang, G., Wang, S., Cao, J., Li, M., Pang, X., et al., 2018. Review of diagenetic facies in tight sandstones: diagenesis, diagenetic minerals, and prediction via well logs. Earth-Science Reviews 185, 234–258.

Lee, S.H., Dutta-Gupta, A., 1999. Electrofacies Characterization and Permeability Predictions in Carbonate Reservoirs: Role of Multivariate Analysis and Nonparametric Regression. SPE56658.

Li, B., Zhang, H., Xia, Q., Peng, J., Guo, Q., 2021. Quantitative evaluation of tight sandstone reservoir based on diagenetic facies -a case of lower silurian kepingtage formation in shuntuoguole low uplift, tarim basin, China. Frontiers in Earth Science 8, 597535.
Li, C., Zhang, X., Zhang, S., 2004. Prediction of a low tight sandstone reservoir in the Majing gas field, sichuan basin, China. Journal of Geophysics and Engineering 1, 228–233.
Lis-Śledziona, A., 2019. Petrophysical rock typing and permeability prediction in tight sandstone reservoir. Acta Geophysica 67, 1895–1911.
Mukherjee, A., Dasgupta, T., Sahoo, A., Srivastava, M., 2017. Delineating Shale Reservoir through Electrofacies Analysis: A New Approach. Conference: AAPG, GTW, MUMBAI, INDIA.
Opuwari, M., Magoba, M., Dominick, N., Waldmann, N., 2021. Delineation of Sandstone Reservoir Flow Zones in the Central Bredasdorp Basin, South Africa, Using Core Samples. Natural Resources Research. https://doi.org/10.1007/s11053-021-09885-7.
Millson, J.A., Quin, J.G., Idiz, E., Turner, P., Al-Harthy, A., 2008. The Khazzan gas accumulation, a giant combination trap in the Cambrian Barik Sandstone Member, Sultanate of Oman: implications for Cambrian petroleum systems and reservoirs. Bulletin of the American Association of Petroleum Geologists 92, 885–917.
Mortensen, J., Engstrøm, F., Lind, I., 1998. The relation among porosity, permeability, and specific surface of chalk from the Gorm field, Danish North Sea. SPE Reservoir Evaluation & Engineering 1 (03), 245–251.
Orsini, C., Rezaee, R., 2012. Depositional systems sequence stratigraphy frameworks & geological modelling of fluvial lbodies. Geological Survey of Western Australia, Report 112, 405.
Pranter, M.J., Ellison, A.I., Cole, R.D., Patterson, P.E., 2007. Analysis and modeling of intermediate-scale reservoir heterogeneity based on a fluvial point-bar out crop analog, Williams fork formation, piceance basin, Colorado. AAPG Bulletin 91, 1025–1051. Available from: https://doi.org/10.1306/02010706102.
Riazi, Z., 2018. Application of integrated rock typing and flow units identification methods for an iranian carbonate reservoir. Journal of Petroleum Science and Engineering 160, 483–497.
Rushing, J.A., Newsham, K.E., Blasingame, T.A., 2008. Rock typing: keys to understanding productivity in tight gas sands. In: SPE Unconventional Reservoirs Conference. Society of Petroleum Engineers.
Serra, O., Abbott, H.T., 1980. The Contribution of logging data to sedimentology and stratigraphy. SPE-9270-PA. https://doi.org/10.2118/9270-PA.
Shanley, K., Cluff, R.M., Robinson, J.W., 2004. Factors controlling prolific gas production from low-permeability sandstone reservoirs: implications for resource assessment, prospect development, and risk analysis. American Association of Petroleum Geologists (AAPG) Bulletin 88 (8), 1083–1121.
Stinco, L., Barredo, S., 2021. Geomechanics and electrofacies characterization of the los molles formation (lower to middle jurassic), neuquén basin. Journal of South American Earth Sciences 110, 103338.
Thomeer, J.H.M., 1960. Introduction of a pore geometrical factor defined by the capillary pressure curve. Journal of Petroleum Technology 12 (3), 73–77.
Tobin, R.C., McClain, T., Lieber, R.B., Ozkan, A., Banfield, L.A., Marchand, A.M.E., et al., 2010. Reservoir quality modeling of tight-gas sands in wamsutter field: integration of diagenesis, petroleum systems, and production data. AAPG Bulletin 94 (8), 1229–1266.

Wang, W., Lin, C., Zhang, X., Dong, C., Ren, L., Lin, J., 2022. Discussion of seismic diagenetic facies of deep reservoir in the east China sea basin. Journal of Petroleum Science and Engineering 208 (Part A), 109352.

Wang, M., Yang, Z., Shui, C., Yu, Z., Wang, Z., Cheng, Y., 2019. Diagenesis and its influence on reservoir quality and oil-water relative permeability: a case study in the yanchang formation chang 8 tight sandstone oil reservoir, ordos basin, China. Open Geosciences 11, 37–47.

Yu, P., 2021. Hydraulic unit classification of un-cored intervals/wells and its influence on the productivity performance. Journal of Petroleum Science and Engineering 197, 107980.

Zhang, X.G., Zhang, T., Ju, C.X., Lin, C.Y., Dong, C.M., Lin, J.L., et al., 2018. A quantitative method and workflow for diagenetic facies prediction of tight reservoir based on multiple scale information, vol. 7. China Patent, ZL201510895321.

Further reading

Abbot, H.T., 1980, The contribution of logging data to sedimentology and stratigraphy, 55th Annual Fall Meeting of AIME (SPE 9270), Dallas, Texas.

Mahjour, S.K., Al-Askari, M.K.G., Masihi, M., 2016. Flow-units verification, using statistical zonation and application of stratigraphic modified lorenz plot in tabnak gas field. Egyptian Journal of Petroleum 25 (2), 215–220. Available from: https://doi.org/10.1016/j.ejpe.2015.05.018.

Rezaee, M.R., Motiei, H., Kazemzadeh, E., 2007. A new method to acquire M exponents and tortuosity factor for microscopically heterogeneous carbonates. Journal of Petroleum Science and Engineering 56 (4), 241–251.

Rushing, J.A., Newsham, K.E., 2001, An integrated work-flow model to characterize unconventional gas resources: Part II. Formation evaluation and reservoir modeling: SPE Paper 71352, Proceedings, Society of Petroleum Engineers Annual Technical Conference and Exhibit ion, CD-ROM, 11 p.

Shirmohamadi, M., Kadkhodaie, A., Rahimpour-Bonab, H., Faraji, M.A., 2017. Seismic velocity deviation log: an effective method for evaluating spatial distribution of reservoir pore types. Journal of Applied Geophysics 139, 223–238.

CHAPTER 4

Pore type and pore size distribution

4.1 Introduction

Pore system properties of tight gas sandstone reservoirs have an essential control on fluid flow and production behavior of the formation. Porosity, permeability, and their relationship in terms of pore network properties (e.g., pore type, pore-throat size, and distribution) are the major elements of the pore system, among which pore-throat size is the main parameter for reservoir evaluation of the low permeability tight sandstones. The distribution of these elements in the reservoir is significantly related to the initial depositional facies, mineralogy, diagenesis, and in situ reservoir conditions (e.g., overburden pressure, water saturation).

The low permeability of tight gas sandstones can directly be attributed to their complicated pore system consisting of small to very small pores and/or tortuous pore throats (Rushing et al., 2008). Pores of conventional reservoirs are fully connected and their porosity is high, whereas pores of tight reservoirs are occluded, and there is less connectivity between them (Rezaee et al., 2012). Naik (2003) described the difference in pore systems between conventional and tight sandstones based on thin-section photos, where pore spaces in conventional sandstones are interconnected, and the gas is able to flow easily from the rock. In contrast, pore spaces in tight sandstones are irregularly distributed, and the connection between them is limited. Photomicrographs of Fig. 4—1 show differences between pore spaces in a conventional sandstone and tight sandstone. In fact, tight sandstones due to the small pore-throat size and very narrow capillaries connecting the pores are characterized by very low permeability, high water saturation, and high capillary pressure, as gas flows through them at low rates (Thomas and Ward, 1972; Naik, 2003; Dutton and Loucks, 2010).

4.2 Pore type in tight sandstone reservoirs

Pore type of reservoir rocks has an essential control on their porosity and permeability relationship and reservoir production. In fact, tight

Reservoir Characterization of Tight Gas Sandstones
DOI: https://doi.org/10.1016/B978-0-323-90180-2.00006-2

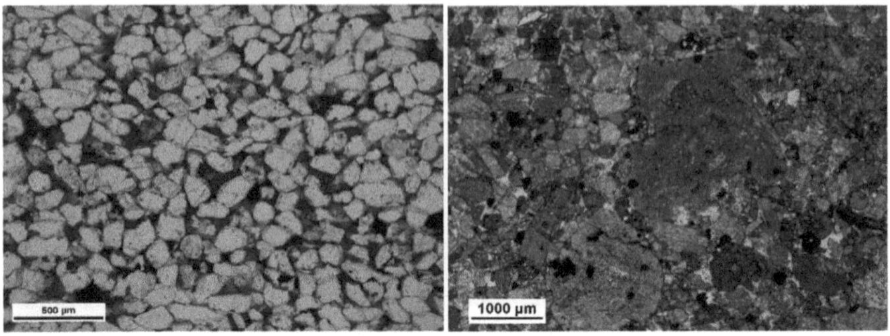

Figure 4–1 Thin-section photomicrographs of a conventional quartz arenite sandstone (Zagros Basin, SW Iran) with a good connection between primary intergranular pores and a tight litharenite sandstone (Moghan Basin, NW Iran) with isolated dissolved pores. Thin sections are blue-dyed (For interpretation of the references to color in this figure legend, the reader is referred to the web version of this article.).

sandstones can be differentiated from the conventional sandstone reservoirs based on the dominant pore types, pore size, and pore connectivity (pore-throat size). Successful exploitation of tight gas sandstone reservoirs requires a basic understanding of pore system properties and the factors controlling those properties (Rushing et al., 2008). In conventional sandstones, permeability is mainly controlled by a well-connected primary, intergranular pore with large pore throats. But in tight sandstones, secondary pore types under the effect of diagenesis, and as a result of partly to complete dissolution of framework grains are dominant. In addition, microporosity related to a matrix, detrital and authigenic clays, and weathered or altered grains (e.g., feldspars and volcanic and mudstone rock fragments) shares some part of pore systems in tight sandstones (Loucks and Dutton, 2007; Chen et al., 2014; Cao et al., 2020). Secondary, dissolution pores and micropores in tight sandstones have small pore throats that less contribute to permeability (Dutton and Loucks, 2010).

There are many different works in the literature focusing on pore systems and pore types of tight gas sandstone reservoirs. Dutton and Loucks (2010), by a study of the lower Tertiary Wilcox sandstones from the Gulf of Mexico Basin, concluded that in addition to the pore volume, pore types show significant change during the burial diagenesis. They showed that the reduction in primary pores volume by compaction and cementation during the burial diagenesis is associated with an increasing proportion of secondary pores and micropores that are preserved at the greater burial depths and temperatures than primary pores. This change in the proportion of pore types as a function of temperature is shown in the triangular plot Fig. 4–2. The results show that many deep types of sandstone at temperatures >150°C are likely to be dominated by these pore types, and in fact they are tight. Similarly, Guo et al. (2020) by casting thin sections and field emission-scanning electron microscopy (FE-SEM), recognized four pore types in the tight reservoir sandstones of Chang 7 unit of the Upper Triassic Yanchang Formation, Ordos Basin of China. They include (1) original intergranular pores with less development and triangular or polygonal shape and diameters less than 30 μm, (2) dissolved intergranular pores with irregular shape and widely distributed in the diameter range of less than 120 μm, (3) dissolved intragranular pores with poor connectivity and a diameter of 1–20 μm, and (4) micropores (generally less than 2 μm) related to the authigenic clay minerals such as chlorites, illites, and mixed Illite/smectite layers.

Moreover, Soeder and Chowdiah (1990) and Dutton and Loucks (2010) classified tight reservoir sandstones, in terms of pore system properties, into three rock types as follows.

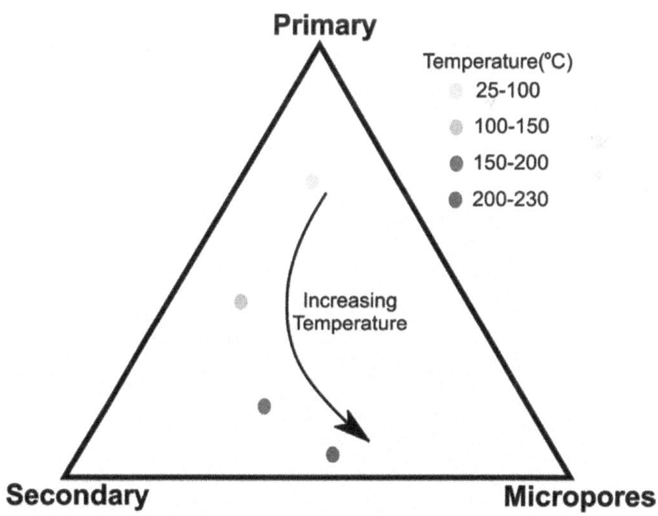

Figure 4–2 Ternary plot showing the average proportions of pore types for different temperature intervals of Wilcox sandstones in the Gulf of Mexico Basin. As is seen, the proportions of secondary pores and micropores toward the greater temperatures are increased. *Modified after Dutton, S.P., Clift, S.J., Hamilton, D.S., Hamlin, H.S., Hentz, T.F., Howard, W.E., et al., 1993. Major low-permeability-sandstone gas reservoirs in the continental United States. In: Bureau of Economic Geology, Report of Investigations No. 211. The University of Texas at Austin, pp. 221.*

1. Sandstones with open intergranular pores and pore throats plugged by authigenic clay minerals, and permeability between 10 and 100 μD (μD = microdarcy = 0.001 md = 10^{-6} Darcy).
2. Sandstones with highly altered primary pores occluded with authigenic cements (quartz or calcite) and permeability between 1 and 10 μD. The majority of pores and pore throats are reduced to narrow slots (Fig. 4–3) connecting significant secondary pores created by grain dissolution (Dutton and Loucks, 2010).
3. Muddy sandstones with intergranular pores filled with detrital clay matrix, porosity is dominated by microporosity and permeability is less than 1 μD.

Mousavi and Bryant (2012) through grain-scale simulation of porosity-reducing mechanisms (ductile grain deformation and cement precipitation) in tight gas sandstones showed that pore space is poorly connected in these rocks. Soeder and Chowdiah (1990) based on the petrographic evidence from the tight sandstones of the Frontier, Mesaverde, and Travis Peak formations in the United States, differentiated them by loss of primary porosity under the effect of diagenesis, development of secondary dissolved pore types and the presence of slot pores between adjoining quartz overgrowths on sand grains. Slot pores act as interconnected pathways for fluid flow. Reservoir characteristics of some tight sandstone formations in different basins of the world have been shown in Table 4–1. Nelson (2009) discussed pore-throat sizes as a continuum from conventional reservoirs to tight gas sandstones and shales. According to this

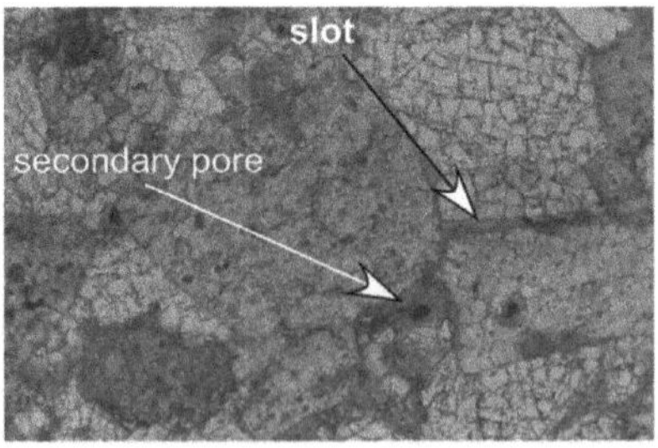

Figure 4–3 Photomicrograph showing slot-type pores and pore throats commonly found in low-permeability reservoirs. The slot-type pore network commonly consists of secondary, solution-derived pores that are connected by narrow, sheet-like slots (**Shanley et al.**, 2004). Frontier Formation, Amoco Shute Creek 1, 10,779.8 ft (3285.6 m), 100x, plane-polarized light. *From Shanley, K.W., Cluff, R.M., Robinson, J.W., 2004. Factors controlling prolific gas production from low-permeability sandstone reservoirs: implications for resource assessment, prospect development, and risk analysis. AAPG Bulletin, 88, 1083–1121.*

Table 4–1 Reservoir characteristics of some tight sandstone formations of the world.

Formation	Porosity (%)	Permeability (μD)	Pore throat Size (μm)
Upper Cretaceous Lance Formation, Greater Green River Basin, Wyoming	4.6–10.4	3–338	0.17–0.89
Upper Cretaceous Mesaverde Formation, Piceance Basin, Colorado	3–11	0.5–9	0.04–0.18
Lower Cretaceous Travis Peak Formation, East Texas	3–6.3	0.09–16.5	0.044–0.22
Upper Jurassic Bossier Interval, East Texas	1–14 (RRT), 1–8 (non-RRT)	0.3–500 (RRT), 0.03–20 (non-RRT)	0.094–1 (RRT), 0.010–0.024 (non-RRT)
Low-Permeability Sandstone Reservoirs of Rocky Mountain Region, western United States	5–10	0.1–10	< 0.5
Cambrian Barik Sandstone Member, Sultanate of Oman	0.5–24.2	0.01–421	–
Permian Shihezi Formation of Bohai Bay Basin, North China	1.2–11.6	$0.01(10^{-3})–0.675(10^{-3})\mu m^{-2}$	2.92
Triassic Chang 8 tight sandstones, Zhenjiang area, Ordos Basin, China	0.3–18.1	< 0.1–9.45	0.0617–0.1947
Upper Triassic Xujiahe Formation, western Sichuan Basin, China	0.4–9.38	$0.01–2034.70 \times 10^{-3}\ \mu m^2$	0.004–0.188
Tight gas sandstones of Western Australia	0.8–20	0.01– < 10	0.1–1

Source: Data and information for the Lance, Mesaverde, Travis Peak, and Bossier formations are from Nelson (2009), tight sandstones of Western Australia from Rezaee et al. (2012), Rocky Mountain Region from Shanley et al. (2004), and Upper Triassic Xujiahe Formation from Yang et al. (2021).

continuum, the pore-throat sizes in tight gas sandstone reservoirs range from about 2 to 0.03 μm. In these reservoirs, productive and nonproductive units are interlayered spatially, and there are overlapping values of their pore and pore-throat size. He also mentioned that the pore-throat spectrum in low permeability rocks, from the economic recovery view, is of significant importance, and pore-throat sizes, instead of porosity and permeability, could be used as a cut-off criterion in determining the net pay of these rocks. Moreover, pore-throat size distribution has important control on irreducible water saturation, threshold gas column, and transition-zone thickness.

4.3 Pore structure characteristics

The pore structure of clastic rocks, which is responsible for a considerable range of static and dynamic reservoir characteristics, is attributed to the various factors related to geological processes including provenance and transport, depositional environment, and diagenesis (Behrenbruch and Biniwale, 2005). Diagenetic processes such as compaction, cementation, and dissolution by qualitative and quantitative change of pore system can affect the pore structure of reservoir rocks. Pore structures of tight sandstones by controlling flow capacity, water saturation, and the producible pore volume have a significant effect on gas production scenarios (Lyu et al., 2021). Thus a deep understanding of the reservoir properties of tight sandstone reservoirs is required for pore geometry analysis (Cluff and Byrnes, 2010). In these sandstones, the pore geometry and structure are mainly different from the conventional reservoir sandstones such that the pore system includes narrower fluid pathways and to some extent closed and disconnected pore spaces (Prodanovic et al., 2013). The pore throat structure of tight reservoirs is characterized via direct observations [e.g., computed tomography (CT) scanner and scanning electron microscope] and indirect measurements (e.g., Mercury injection, gas adsorption, and nuclear magnetic resonance). Many studies have recognized that the geometry, aperture, distribution, and connectivity of pore throats can be described by pore structure parameters derived from the mercury injection test (Purcell, 1949; Van Brakel et al., 1981). Capillary pressure as a prevailing control on pore-scale fluid flow in tight reservoirs is controlled by pore geometry and structure. Also, in most unconventional gas reservoirs (coals, shales, and tight gas reservoirs), non-Darcy flow effects (slippage, transitional flow, and diffusion) related to pore structure (e.g., ultrafine nature of matrix porosity) are important under certain PVTx conditions and must be considered for measurements using gases (Civan, 2010, Clarkson et al., 2012).

4.4 Pore system analysis in tight gas sandstone reservoirs

The pore system of tight gas sandstones is comprised of a complicated and heterogeneous pore network consisting of poorly connected pores and small pore throats. The presence of authigenic cements such as silica, carbonate, and clays within a compacted fabric has a significant control on pore network connectivity and continuity in these reservoirs. Bustin et al. (2008b) have introduced different methods, classified as radiation and fluid invasion (penetration) methods, to estimate porosity and pore size distribution (PSD) in unconventional gas reservoirs (Fig. 4–4). In radiation methods, X-ray or neutron scattering patterns are used for the interpretation of pore structure. In fluid invasion methods, the structure of

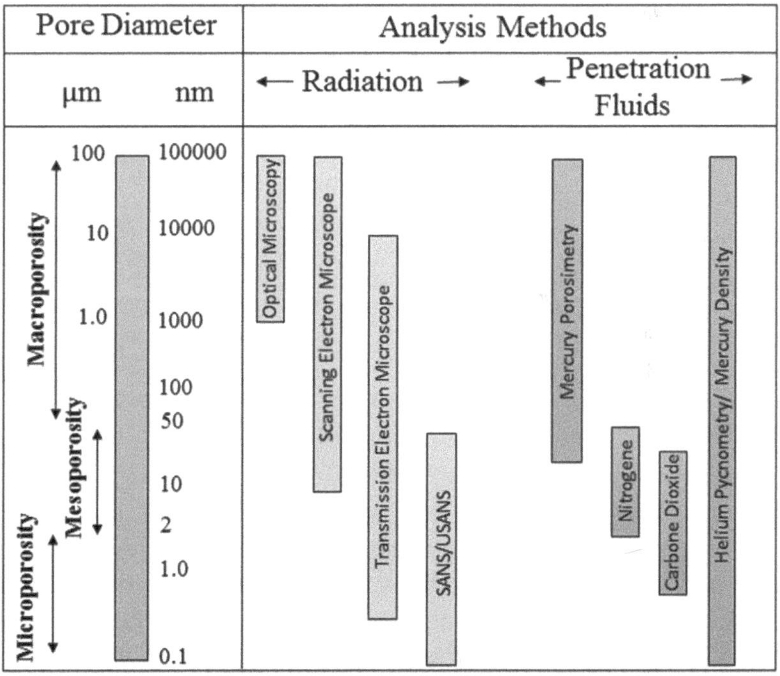

Figure 4–4 Analysis methods used in the estimation of porosity and pore size distribution in unconventional gas reservoirs. *SANS*, Small-angle neutron, *USANS*, ultrasmall-angle neutron. *Modified after Bustin, R.M., Bustin, A., Ross, D., Chalmers, G., Murthy, V., Laxmi, C., et al., 2008b. Shale gas opportunities and challenges. In: Search and Discovery Articles no. 40382 (2009). Adapted from oral presentation at AAPG Annual Convention, San Antonio, Texas, 20–23 April and Clarkson, C.R., Jensen, J.L., Chipperfield, S., 2012. Unconventional gas reservoir evaluation: what do we have to consider? Journal of Natural Gas Science and Engineering 8, 9–33.*

connected pores is investigated by a molecular probe. Therefore the investigation of pore systems of tight sandstone reservoirs requires a quantitative and qualitative analysis using a combination of techniques such as thin-section petrography, SEM, X-ray CT, mercury injection capillary pressure (MICP) tests, nuclear magnetic resonance (NMR) spectroscopy, nitrogen adsorption tests and sonic velocities, which are described in next.

4.4.1 Thin-section petrography

Pore spaces as a component of the rock volume can be observed in detail under the polarizing microscope, especially when thin sections are blue-dyed. By this method, primary and secondary pore types, depositional and diagenetic factors controlling development and distribution in tight sandstone reservoirs can be identified. Although this method provides direct and detailed qualitative information on pore spaces, the quantitative investigation is based on the pore types (mostly macropores) observed under the microscope. Therefore most micropores related to matrix or clay minerals are not recognizable in this resolution. Also, thin sections provide a 2D view of reservoir rock that covers only one line of a 3D volume of the whole rock. Therefore, in tight gas sandstone reservoirs with a heterogeneous and complex pore network, the results of thin-section studies should be controlled by high-resolution techniques such as SEM and NMR.

4.4.2 Scanning electron microscopy and X-ray computed tomography

Pore system analysis by SEM, in addition to the pore type identification, provides information regarding pore geometry, pore size, and pore volume (Cai et al., 2022). Clay minerals are important authigenic components in tight gas sandstones that their type, structure, and distribution are specified by SEM analysis. Many studies of tight gas sandstone reservoirs have considered SEM analysis results as an effective tool to confirm and interpret the results of quantitative analysis of pore systems derived from routine and special core analysis (CCAL and SCAL), MICP, and NMR, and also to describe the main diagenetic features affecting the pore system properties of tight reservoirs (e.g., Zhang et al., 2009; Wang et al., 2015; Lai et al., 2018; Ren et al., 2019; Hong et al., 2020). Due to the high resolution of SEM images, a detailed description of pore morphology is feasible by this method (Su et al., 2022). Wu et al. (2017) by SEM analysis of tight sandstones of Ordos Basin in China identified four pore types, including residual interparticle, intercrystalline, dissolution pores, and small microcracks.

X-ray CT and at a much higher resolution, X-ray microtomography (micro-CT) has widely been used in pore network analysis of tight gas sandstones (e.g., Golab et al., 2010; Liu et al., 2017, Munawar et al., 2018). This method through X-ray tomography over a continuous length scale of rocks provides a three-dimensional nondestructive image of rock samples. The resultant 3D high-resolution images can be used for the analysis of pore structure and investigation of petrophysical parameters such as porosity, permeability, formation factor, and capillary pressure. Moreover, useful quantitative information on two and three-phase flow can be derived from these images (Munawar et al., 2018; Wu et al., 2019; Su et al., 2022).

4.4.3 Mercury injection capillary pressure tests

Capillary-pressure data are considered an effective tool in evaluating the producibility of tight gas sandstones (Mousavi and Bryant, 2012). MICP curves are generally used for quantitative analysis of reservoir pore structure. In conventional reservoirs, the Winland Eq. (4.1), published by Kolodzie (1980), shows the relationship between pore throat size (r, μ, m), porosity (φ, %) and air permeability (k, md) corresponding to a mercury saturation of 35% (r_{35}). Later, the Winland equation was extended and modified by other researchers such as Pittman (1992), Eq. (4.2) and Rezaee et al. (2006), Eq. (4.3)

$$\mathrm{Logr35} = 0.732 + 0.588\mathrm{Logkair} - 0.864\mathrm{Log}\varphi \tag{4.1}$$

$$\mathrm{Logk} = -1.221 + 1.415\mathrm{Log}\varphi - 1.512\mathrm{Logr25} \tag{4.2}$$

$$\mathrm{Logk} = -1.160 + 1.780\mathrm{Log}\varphi - 0.930\mathrm{Logr50} \tag{4.3}$$

The application of these equations in tight gas sandstone reservoirs is problematic due to the complex pore structure consisting of poorly connected pores and small pore throat sizes. Comisky et al. (2007), by evaluating 63 core samples obtained from 9 tight gas sands in 6 basins of the US and Argentina, provided a comparison of the applicability of several capillary pressure—based empirical models to estimate absolute permeability (Table 4—2). They concluded that the current models have been developed based on

Table 4–2 A summary of capillary pressure–based empirical models for estimation of permeability.

Method	Year	Number of samples	Sample source	Permeability measurement type	Net confining pressure(psi)	Permeability range (md)	Porosity range (dec)
(Purcell, 1949)	1949	27	Sandstone	Air	Ambient	0.1–1459	Unknown
Winland	1980	322	Total	Air, SS Klinkenberg	Ambient	Unknown	Unknown
		56	Sandstone	SS Klinkenberg			
		26	Carbonate	SS Klinkenberg			
		240	Various	Air			
Swanson Air	1981	319	Total	Air	Ambient	0.02–5000[a]	Unknown
		203	Sandstone				
		116	Carbonate				
Swanson Brine	1981	39	Sandstone	Brine	1000	0.02–1000[b]	Unknown
Walls–Amaefule	1985	35	Sands	Pulse decay	3000–4000	0.00002–70	0.03–0.16
Katz–Thompson	1986	50	Total	SS Klinkenberg	Ambient	0.005–5000	Unknown
Pittman	1992	202[c]	Sandstone	Air	Ambient	0.05–1000	0.03–0.28
(Kamath, 1992)	1992	454	Total	Air, pulse decay	Ambient	0.00002–2000	0.03 unknown[d]
		158	Chevron	Air			
		268	Swanson	Air			
		29	Walls–Amaefule	Pulse decay	3000–4000		
Huet–Blasingame	2005	89	Sandstone	SS Klinkenberg	800	0.0041–8340	0.03–0.34
OU (Dastidar)	2007	150	Sands	USS Klinkenberg (CMS 300)	800	0.0001–10000	0.005–0.30
Comisky et al. (2007)	2007	63	Sands	SS Klinkenberg	800	0.0001–0.2	0.02–0.15

[a]Estimated from Swanson (1981).
[b]Estimated from Swanson (1981).
[c]103 of the 202 samples taken from the Winland data set.
[d]Upper range of porosity for the Chevron dataset not provided in Kamath (1992).
Source: Comisky, J.T., Newsham, K.E., Rushing, J.A., 2007. A comparative study of capillary-pressure-based empirical models for estimating absolute permeability in tight gas sands. In: SPE Annual Technical Conference and Exhibition held in Anaheim, California, U.S.A., 11–14 November, 2007.

datasets with a permeability greater than 1md, and also these models suffer from the lack of a consistent reference permeability. They also mentioned that correlations should use the Klinkenberg-corrected permeability as the standard reference, especially for tight gas sands.

Chao-Liu et al. (2010) via integrating porosity (φ), the maximum radius of connected pore throats (Rm or PD), and sorting degree (Sp), proposed a new petrophysical parameter δ, (Eqs. 4.4 and 4.5) for estimation of permeability and electrical parameters (i.e., Formation factor, F) in tight sandstones.

$$\delta = \frac{\varphi^c R_m}{\sqrt{Sp}} \tag{4.4}$$

$$\delta = \frac{\varphi^c}{PD \times \sqrt{Sp}} \tag{4.5}$$

In these equations, c is a dimensionless constant, φ is derived from the conventional logs or NMR log, and PD is determined from the capillary pressure curve derived from the T_2 distribution. Sorting degree, Sp, is defined as Eq. (4.6).

$$Sp = 1 - \frac{1}{N}\sum_{i=1}^{N} \frac{(\overline{\mathbf{x}})^4}{\left(\left(\overline{\mathbf{x}}^2 + (x_i - \overline{\mathbf{x}})^2\right)\right)^2} \tag{4.6}$$

where N is the total number of pores, x_i is the individual pore size and x is the average of all pore sizes.

Rezaee et al. (2012) by mercury injection tests on 16 core plugs from one of the Western Australian tight gas sand fields, showed that the equations mentioned previously Eqs. (4.1)–(4.3) are not applicable for permeability estimation in tight gas reservoirs. Therefore they developed empirical equations through multiregression analysis for calculating permeability from porosity and pore throat radius at several mercury saturation percentiles (5%–80%). According to their study results, pore throat radii corresponding to 10% mercury saturation (r_{10}) yield the best correlation with permeability. Consequently, they introduced a new Eq. (4.7) for tight gas sandstones, based on which the dominant pore throat radius was determined to be 0.1–1 μm. They attributed this range of pore sizes to micropores and remnant intergranular mesopores (Fig. 4–5 and Fig. 4–6).

$$\text{Logk} = -1.92 + 0.949\text{Log}\varphi + 2.18\text{Logr}10 \tag{4.7}$$

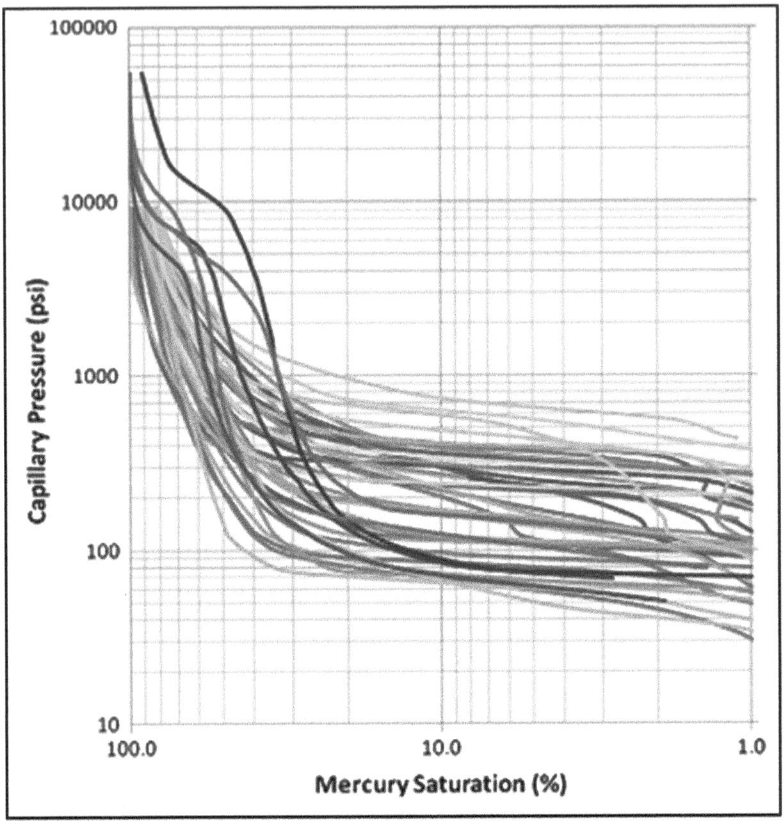

Figure 4–5 Capillary pressure vs mercury saturation for core plug samples from tight gas sandstones of Western Australia. *From Rezaee, R., Saeedi, A., Clennell, B., 2012. Tight gas sands permeability estimation from mercury injection capillary pressure and nuclear magnetic resonance data. Journal of Petroleum Science and Engineering 88–89, 92–99.*

Similarly, Sujuan et al. (2011) by correlating porosity, permeability, and pore throat radii corresponding to different mercury saturation of the tight sandstones of the Upper Triassic Xujiahe Formation concluded that the pore throat radius at 20% mercury saturation (R_{20}) is the best permeability predictor. As a result, they identified four petrophysical rock types and mentioned that R_{20} values are intimately related to the depositional texture (grain size and sorting), detrital composition, and authigenic minerals. Thus it can be concluded that the dominant pore throat radius controlling the permeability in tight sandstones generally corresponds to lower mercury saturation values than that of conventional reservoirs. In

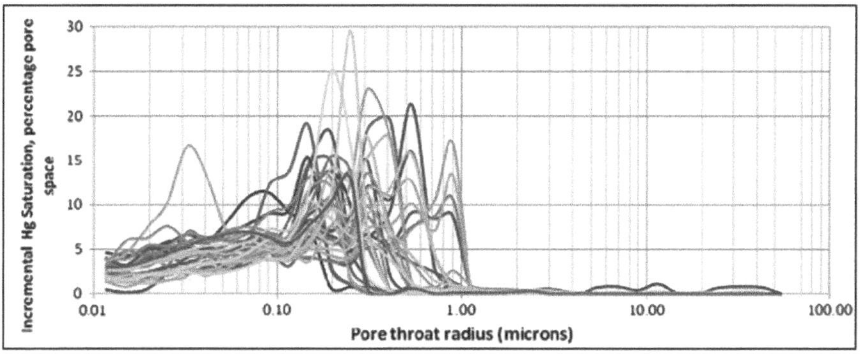

Figure 4–6 Pore size distribution diagram showing the dominant pore throat sizes range between 0.1 and 1 μm for core plug samples from tight gas sandstones of Western Australia. *From Rezaee, R., Saeedi, A., Clennell, B., 2012. Tight gas sands permeability estimation from mercury injection capillary pressure and nuclear magnetic resonance data. Journal of Petroleum Science and Engineering 88–89, 92–99.*

higher mercury saturation, pore throat radius indicates nanopores that seem have little contribution to permeability (Rezaee et al., 2012).

Due to the limitation of MICP tests, in the detection of small and disconnected pores, the high-pressure mercury intrusion (HPMI) technique which is an extension of the conventional mercury intrusion test, was developed. On the other hand, as extremely high-pressure tests of HPMI may damage the sample and cause some artificial cracks, rate-controlled mercury intrusion (RCMI) with a very low rate of mercury injection and no change in interfacial tension and contact angle was developed.

Also, Ziarani and Aguilera (2012) proposed a new model, derived from formation resistivity factor (*F*) and permeability (*K*, md) data, for estimation of pore throat radius (r_T, μm) of tight gas sandstone reservoirs Eq. (4.8). Consequently, they determined pore throat radii between 0.01 and 5 μm (mostly between 0.01 and 1 μm) for tight sandstones of Mesaverde Formation.

$$rT = a(FK)b + c \tag{4.8}$$

In this equation, constant *a* varies between 0.024 and 0.11, exponent *b* varies between 0.51 and 0.57, and constant *c* varies between 0.001 and 0.015.

4.4.3.1 Capillary pressure and pore size distribution

Capillary pressure is defined as the pressure difference across the interface between two immiscible fluids. Capillary pressure that has an inverse relationship with pore-throat size Eq. (4.9), Washburn (1921) has been used for the characterization of pore-throat size distribution.

$$P_c = \frac{2\gamma Cos\theta}{r_c} \tag{4.9}$$

where P_c is capillary pressure (psi), γ is the interfacial tension (dynes/cm), θ is the contact angle, and r_c is the pore throat radius (μm).

PSD in unconventional reservoirs is applicable in the evaluation of gas storage and transport mechanism and also identification of dominant pore throat radii controlling fluid flow (Clarkson et al., 2012). Thus a size scale of pore-throat is better than the permeability scale when gas charging is considered in low permeability reservoirs. Tight gas sandstones due to their small pore-throat size are characterized by high water saturation and high capillary pressure (Dutton and Loucks, 2010). In these reservoirs, water is held by significant capillary pressure forces, and thus the gas flow through small throats is blocked (Cluff and Byrnes, 2010; Bahrami et al., 2012). Amann-Hildenbrand et al. (2012) introduced the capillary-controlled viscous flow as the prominent transport process of tight gas sandstones, and a much smaller extent of water drainage from their pore system compared to the conventional reservoirs.

4.4.4 Nuclear magnetic resonance spectroscopy

NMR spectrometry is related to the response of hydrogen protons in a magnetic field. The magnetic interactions between protons are reflected as longitudinal relaxation (T_1) and transverse relaxation (T_2). The amplitude of the NMR signal is proportional to the density of proton spin by which porosity is estimated (Howard and Kenyon, 1992; Wang et al., 2015). NMR is the only valuable log that provides information on the PSD and pore structure. This information despite the laboratory tests such as mercury intrusion, SEM, and thin sections that are limited to core samples and plugs, is recorded as a continuous log in depth (Chao-Liu et al., 2010). Rezaee et al. (2012) conducted an NMR analysis on tight gas sandstones of Western Australia, as shown in Fig. 4–7, for the study of PSD and permeability estimation. They concluded that due to the lack of macropores, T_2 times for tight gas sands are shorter than that observed in normal sands. They also used Eq. (4.10), derived from a multiregression analysis using porosity and T_2 peak, to estimate permeability with high accuracy.

$$K = -0.0461 - 0.0601\,T2peak + 4.37\varphi \tag{4.10}$$

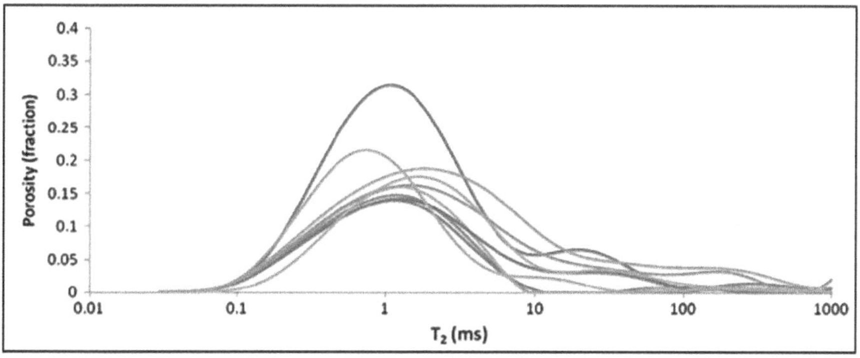

Figure 4–7 Nuclear magnetic resonance test results for tight gas sandstones of Western Australia. *From Rezaee, R., Saeedi, A., Clennell, B., 2012. Tight gas sands permeability estimation from mercury injection capillary pressure and nuclear magnetic resonance data. Journal of Petroleum Science and Engineering 88–89, 92–99.*

In this equation, k is permeability (md), φ is fractional porosity, and T_{2peak} is dominant T_2 on T_2 spectrum (ms).

Moreover, Wang et al. (2020) through T_2 distribution analysis of NMR on the Permian tight sandstones in the north Ordos Basin of China, indicated that the clay-rich and fine-grained facies have a high volume of smaller pore throats with capillary and clay-bound water in their micropores.

4.4.5 Nitrogen adsorption tests

Micropores and nanopores contribute to a significant part of pore volume in tight sandstones. The importance of these pore types on pore system heterogeneity, gas storage, and flow capacity in unconventional reservoirs has widely been mentioned in the literature (e.g., Wu et al., 2017; Lai et al., 2019; Lyu et al., 2021). The isothermal adsorption of nitrogen at low temperature and low pressure is used to evaluate the pore-throat size distribution, specific surface area, pore volume distribution, and also prediction of shape characteristics of pore throats by isothermal adsorption and desorption curves (Zhao et al., 2019). The range of pore sizes measured by this method is 1.7–193.5 nm.

4.4.6 Velocity deviation analysis

Complexity in the pore system of tight gas sandstone reservoirs could be reflected as variations in sonic velocity. In other words, quantitative and qualitative changes in pore systems of the rock show an intimate relationship with velocity as is indicated by the Wyllie time-average [Eq. (4.11), Wyllie, et al., 1956]. Anselmetti and Eberli (1999) developed this equation as a velocity deviation log (VDL) Eq. (4.13) for pore type analysis of carbonate reservoirs. On the other hand, Kadkhodaie et al. (2021) extended the VDL concept in combination with seismic attributes to unravel the pore system heterogeneity and interpret reservoir rock types of tight gas sandstones of the Whicher Range field in the Perth Basin of Western Australia. Consequently, they correlated variations in this log with depositional facies and diagenetic features controlling the dominant pore types of reservoir rocks.

$$\frac{1}{V_{rock}} = \frac{(1-\varphi)}{V_{matrix}} + \frac{\varphi}{V_{fluid}} \tag{4.11}$$

In this equation, φ is fractional porosity, V_{rock}, V_{matrix}, and V_{fluid} are compressional velocity (ft/sec) of the formation, rock matrix, and interstitial fluids, respectively. Sonic porosity (φ_s) can be derived from Eq. (4.11), as indicated by Eq. (4.12).

$$\phi_S = \frac{(DT - DT_{matrix})}{\left(DT_{fluid - DT_{matrix}}\right)} \tag{4.12}$$

Where DT is the transit time of the wave through the formation derived from the log, and DT_{fluid} and DT_{matrix} are transition times of the wave through the rock matrix and pore-filling fluid, respectively. Accordingly, the VDL is obtained by solving Eq. (4.13) as the difference between real velocity (Vp_{real}) derived from the sonic log and synthetic velocity (Vp_{syn}) derived from the calculated sonic log (DT_{syn}). DT_{syn} is determined by Eq. (4.12), for which the porosity from the neutron or density logs is replaced by the sonic porosity (φ_s). Also, the matrix velocity (DT_{matrix}) is based on the dominant lithology of the reservoir, and the velocity for the water-based fluid by default is set to 189.5 (μs/*ft*).

$$VDL = V_{P_{real}} - V_{P_{syn}} = \left(\frac{304.8}{DT_{log}} - \frac{304.8}{DT_{syn}}\right) \times 1000 \tag{4.13}$$

4.4.7 Integrated pore characterization

In unconventional reservoirs and especially tight gas sandstones, due to the highly heterogeneous nature and a wide range of pore sizes (nano to micro and macro-scaled pore structures), PSD analysis is not limited to a specific method or technique. In this respect, the combined use of various techniques, as has been mentioned by many researchers in the literature, provides comprehensive information on the pore system properties of these reservoirs. This is because any method individually has limitations due to the effective ranges of measurements (Xiao et al., 2020), and so with the combined approach, the shortcomings of each method are compensated by other methods. For example, Su et al. (2022) considered a combination method using SEM, MICP, and micro-CT for the investigation of the pore structure of tight sandstones from the Lucaogou Formation of Junggar Basin in Northwestern China. Their study results showed tens of nanometers to tens of micrometers for the pore size of the sandstones. They concluded that the pore throats with a diameter larger than 2 μm have the dominant control on the flow potential of the reservoir. Similarly, Wu et al. (2019) utilized various methods, including SEM, MICP, NMR, and X-ray CT tests for the study of geometrical, topological, and fractal characteristics of tight sandstones from the Shahejie Formation in a few basins of China. By comparing the results of these methods, they concluded that NMR and MICP cover a wider range of pore sizes (ultramicropores to macropores) than SEM and X-ray CT, and thus these methods can be used for the analysis of the fractal dimension of all pore types. They also mentioned that while X-ray CT is a promising technique that can measure more geometry and topology properties such as connectivity, coordination number, pore and throat shape factor, and flow properties of the pore spaces, it cannot capture sub-resolution micropores. Therefore this technique is not suitable for rocks with high heterogeneity and complex pore systems such as tight reservoirs.

4.5 Wettability

Small pore-throat size in low permeability (tight) sandstone reservoirs results in high water saturation and high capillary pressure (Thomas and Ward, 1972; Dutton and Loucks, 2010). Wettability has substantial control on capillary forces and it affects the in situ water saturation, transport porosity, and effective phase permeability (Amann-Hildenbrand et al., 2012). Moreover, the significant influence of water saturation on the effective gas permeability makes it a critical parameter in reservoir characterization of tight sandstone reservoirs (Amiri et al., 2012; Clarkson et al., 2012).

4.6 The effect of overburden stress on porosity and permeability in tight reservoirs

In tight reservoirs, there is a difference between permeability measured at routine conditions and that measured at confining stress (reservoir condition), and this difference becomes more significant with decreasing permeability and increasing confining pressure (Walls et al., 1982; Wei et al., 1986; Byrnes, 1997; Castle and Byrnes, 1998; Shanley et al., 2004; Cluff and Byrnes, 2010). Investigation of confining pressure on porosity and permeability in both tight gas sandstones and shale shows that the effect of overburden stress on permeability values in low-permeability sandstone reservoirs is more significant than that is observed on porosity values (e.g., Byrnes et al., 1979; Bowers and Katsube, 2002; Nelson, 2009). In fact, the porosity response to increasing overburden pressure is small, and this is mostly attributed to well-cemented fabric and rigid framework in tight sandstones (Shanley et al., 2004). However, there are many references in the literature on tight gas sandstones indicating the significant decrease of permeability with increasing overburden or confining pressure (e.g., Thomas and Ward, 1972; Byrnes, 1997; Jones and Owens, 1980; Byrnes, 2003; Dutton and Loucks, 2010). According to the study results of Davies and Davies (1999), the reduction rate of permeability with increasing stress, in unconsolidated sandstones depends on porosity and permeability volume, and in sandstones with the highest porosity and permeability, this reduction rate is the greatest. In contrast, in the low permeability sandstones, the reduction rate of permeability as a function of increasing stress depends on the pore type such that in rocks with the dominance of slot pores in their system, it is the greatest. According to Byrnes (2005), pore throats in tight sandstones could decrease by 50%–70% with increasing overburden stress. The main reason for dramatic decrease in permeability as a function of increasing overburden stress is attributed to the dominance of slot pores and pore throats in pore systems of tight sandstones and the fact that their permeability is very sensitive to reduction in the pore throat size (Davies and Davies 1999; Rezaee et al., 2012). Through the study of tight gas sandstones of Western Australia, Rezaee et al. (2012) showed that the reduction in porosity with increasing confining pressure is not significant (Fig. 4–8A), but permeability values are reduced significantly with increasing the confining stress such that it changes from about 0.1 mD to less than 0.001 mD at 5000 psi (Fig. 4–8B).

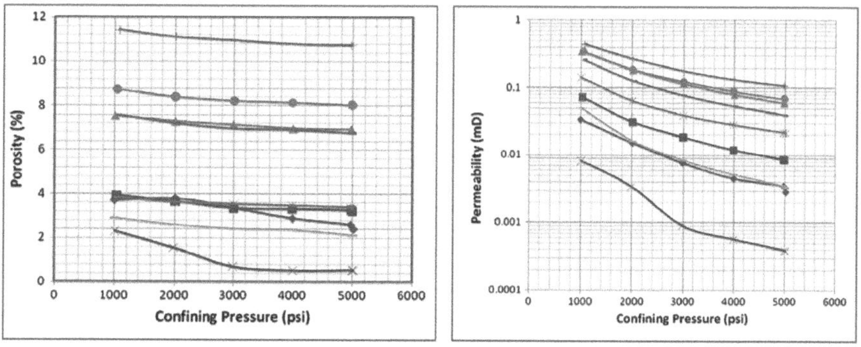

Figure 4–8 The effect of confining pressure on core samples porosity and permeability in different pressures in tight gas sandstones of Western Australia. *From Rezaee, R., Saeedi, A., Clennell, B., 2012. Tight gas sands permeability estimation from mercury injection capillary pressure and nuclear magnetic resonance data. Journal of Petroleum Science and Engineering 88–89, 92–99.*

4.7 Pore size reduction by diagenetic processes

What makes tight sandstones different from the conventional reservoirs is the severity of the diagenetic processes damaging reservoir quality. At the time of deposition, there were no tight sandstones and almost all sandstone formations were classified as conventional reservoirs. Due to the deep burial and severe diagenetic overprints, reservoir sandstones lose much of their porosity, permeability, and other reservoir properties. Physical and compaction along with the cementation reduce the macropores through filling and converting them into micropores and slop pores. Clay minerals in the form of cements and matrix act as pore-filling (e.g., kaolinite), pore-lining (e.g., chlorite), and pore bridging (e.g., illite) causing a severe reduction of reservoir porosity and permeability.

4.8 Pore size reduction by formation damage

In tight reservoirs, in addition to the primary nature of matrix and diagenesis that control the pore system properties, formation damage during drilling, completion, and stimulation operations can significantly affect the reservoir behavior and production in the field. One of the most important and primary agents of near-wellbore formation damage affecting the performance and the rate of productivity in these reservoirs is phase trapping and water blocking, as has widely been mentioned in the literature (e.g., Jamaluddin et al., 1998; Davies and Davies, 1999; Bennion et al., 2006;

Rahman et al., 2007). Bahrami et al. (2012) by investigating of tight reservoir sandstones of the Perth Basin in Western Australia highlighted water blocking as a major damage mechanism, especially in water-sensitive tight sand formations.

References

Amann-Hildenbrand, A., Ghanizadeh, A., Krooss, B.M., 2012. Transport properties of unconventional gas systems. Marine and Petroleum Geology 31, 90–99.

Amiri, M., Yunan, M.H., Zahedi, G., Jaafar, M.Z., Oyinloye, E.O., 2012. Introducing new method to improve log derived saturation estimation in tight shaly sandstones—a case study from Mesaverde tight gas reservoir. Journal of Petroleum Science and Engineering 92–93, 132–142.

Anselmetti, F.S., Eberli, G.P., 1999. The velocity-deviation log: a tool to predict pore type and permeability trends in carbonate drill holes from sonic and porosity or density logs. AAPG Bulletin 83 (3), 450–466.

Bahrami, H., Rezaee, R., Clennell, B., 2012. Water blocking damage in hydraulically fractured tight sand gas reservoirs: an example from Perth Basin, Western Australia. Journal of Petroleum Science and Engineering 88–89, 100–106.

Behrenbruch, P., Biniwale, S., 2005. Characterisation of clastic depositional environments and rock pore structures using the Carman-Kozeny equation: Australian sedimentary basins. Journal of Petroleum Science and Engineering 47, 175–196.

Bennion, D., Thomas, F., Schulmeister, B., Romanova, U., 2006. Water and oil base fluid retention in low permeability porous media—an update. In: Paper PETSOC-2006-136 Presented at the Canadian International Petroleum Conference. Calgary, Alberta, Canada, 13–15 June. Available from https://doi.org/10.2118/2006-136.

Bowers, G.L., Katsube, T.J., 2002. The role of shale pore structure on the sensitivity of wire-line logs to overpressure. In: Huffman, A., Bowers, G. (Eds.), Pressure regimes in sedimentary basins and their prediction: AAPG Memoir 76, 43–60.

Bustin, R.M., Bustin, A., Ross, D., Chalmers, G., Murthy, V., Laxmi, C., et al., 2008b. Shale gas opportunities and challenges. In: Search and Discovery Articles no. 40382 (2009). Adapted from oral presentation at AAPG Annual Convention, San Antonio, Texas, 20–23 April.

Byrnes, A.P., 2003. Aspects of permeability, capillary pressure, and relative permeability properties and distribution in low-permeability rocks important to evaluation, damage, and stimulation. In: Proceedings Rocky Mountain Association of Geologists—Petroleum Systems and Reservoirs of Southwest Wyoming Symposium, Denver, Colorado, p 12.

Byrnes, A.P., 1997. Reservoir characteristics of low-permeability sandstones in the Rocky MountainsThe Mountain Geologist, 34(1), 39–51.

Byrnes, A.P., Sampath, K., Randolph, P., 1979. Effect of pressure and water saturation on permeability of western tight sandstones. In Proceedings of the 5th Annual DOE Symposium on Enhanced Oil and Gas Recovery and Improved Drilling Technology. Elsevier Science, Tulsa, Okla, USA, pp. 22–24.

Cai, J., Tian, Z., Wood, D.A., 2022. Pore-Scale Characterization and Fractal Analysis for Gas Migration Mechanisms in Shale Gas Reservoirs. Gulf Professional Publishing, Chapter One.

Cao, B., Luo, X., Zhang, L., Lei, Y., Zhou, J., 2020. Petrofacies prediction and 3-D geological model in tight gas sandstone reservoirs by integration of well logs and geostatistical modeling. Marine and Petroleum Geology 114, 104202.

Castle, J.W., Byrnes, A.P., 1998. Petrophysics of low-permeability Medina Sandstone, northwestern Pennsylvania, Appalachian Basin: paper SPWLA 1998-v39n4a3. The Log Analyst 39 (4), 36–46.

Chao-Liu, L., Can-Can, Z., Xia, L., Fa-Long, H., Li, Z., Wei-Jun, W., 2010. A novel model for assessing the pore structure of tight sands and its application. Applied Geophysics 7 (3), 283–291.

Chen, D., Pang, X., Xiong, L., Wang, L., Xie, M., 2014. Porosity evolution in the tight gas sands of the upper Triassic Xujiahe formation, western Sichuan basin, China. Revista Mexicana de Ciencias Geológicas 31 (3), 361–375.

Civan, F., 2010. Effective correlation of apparent gas permeability in tight porous Media. Transport in Porous Media 82, 375–384.

Clarkson, C.R., Jensen, J.L., Chipperfield, S., 2012. Unconventional gas reservoir evaluation: what do we have to consider? Journal of Natural Gas Science and Engineering 8, 9–33.

Cluff, R.M., Byrnes, A.P., 2010. Relative permeability in tight gas sandstone reservoirs-the "permeability jail" model. In: SPWLA 51st Annual Logging Symposium (SPWLA-2010–58470), OnePetro, 19–23 June.

Comisky, J.T., Newsham, K.E., Rushing, J.A., 2007. A comparative study of capillary-pressure-based empirical models for estimating absolute permeability in tight gas sands. In: SPE Annual Technical Conference and Exhibition held in Anaheim, California, U.S.A., 11–14 November, 2007.

Davies, J.P., Davies, D.K., 1999. Stress-dependent permeability: characterization and modeling. In Proceedings SPE Annual Technical Conf and Exhibition. Available from https://doi.org/10.2118/56813-MS.

Dutton, S.P., Loucks, R.G., 2010. Reprint of: Diagenetic controls on evolution of porosity and permeability in lower Tertiary Wilcox sandstones from shallow to ultradeep (200-6700 m) burial, Gulf of Mexico Basin, U.S.A. Marine and Petroleum Geology 27 (8), 1775–1787.

Golab, A.N., Knackstedt, M.A., Averdunk, H., Senden, T., Butcher, A.R., Jaime, P., 2010. 3D porosity and mineralogy characterization in tight gas sandstones. Lead Edge 29 (12), 1476–1483.

Guo, R., Xie, Q., Qu, X., Chu, M., Li, S., Ma, D., et al., 2020. Fractal characteristics of pore-throat structure and permeability estimation of tight sandstone reservoirs: a case study of Chang 7 of the upper Triassic Yanchang formation in Longdong area, Ordos Basin, China. Journal of Petroleum Science and Engineering 184, 106555.

Hong, D., Cao, J., Wu, T., Dang, S., Hu, W., Yao, S., 2020. Authigenic clay minerals and calcite dissolution influence reservoir quality in tight sandstones: insights from the central Junggar Basin, NW China. Energy Geoscience 1, 8–19.

Howard, J.J., Kenyon, W.E., 1992. Determination of pore size distribution in sedimentary rocks by proton nuclear magnetic resonance. Marine and Petroleum Geology 9, 139–145.

Kolodzie Jr. S., 1980. Analysis of pore throat size and use of the Waxman–Smits equation to determine OOIP in spindle field. In: Paper SPE 9382 Presented at the 1980 Annual Fall Technical Conference of Society of Petroleum Engineers, Colorado, 21–24 September.

Jamaluddin, A.K.M., Bennion, D.B., Thomas, F.B., Ma, T., 1998. Application of heat treatment to enhance permeability in tight gas reservoirs. In: Annual Technical Meeting, 8–10 June, Petroleum Society of Canada, Calgary, Alberta. Available from: https://doi.org/10.2118/98-01.

Jones, F.O., Owens, W.W., 1980. A laboratory study of low-permeability gas sands: paper SPE 7551-PA. Journal of Petroleum Technology 32 (9), 1631–1640.

Kadkhodaie, R., Kadkhodaie, A., Rezaee, R., 2021. Study of pore system properties of tight gas sandstones based on analysis of the seismically derived velocity deviation log: a case study from the Perth Basin of western Australia. Journal of Petroleum Science and Engineering 196, 108077.

Kamath, J., 1992. Evaluation of accuracy of estimating air permeability from mercury-injection data. SPE Formation Evaluation 7 (7), 304–310.

Lai, F.P., Li, Z.P., Zhang, T.T., Zhou, A.Q., Gong, B., 2019. Characteristics of microscopic pore structure and its influence on spontaneous imbibition of tight gas reservoir in the Ordos Basin, China. Journal of Petroleum Science and Engineering 172, 23–31.

Lai, J., Wang, G., Wang, S., Cao, J., Li, M., Pang, X., et al., 2018. Review of diagenetic facies in tight sandstones: diagenesis, diagenetic minerals, and prediction via well logs. Earth-Science Reviews 185, 234–258.

Liu, X., Wang, J., Ge, L., Hu, F., Li, C., Li, X., et al., 2017. Porescale characterization of tight sandstone in Yanchang formation Ordos Basin China using micro-CT and SEM imaging from nm- to cm-scale. Fuel 209, 254–264.

Loucks, R.G., Dutton, S.P., 2007. Importance of micropores in deeply buried Tertiary sandstones along the Texas Gulf Coast. American Association of Petroleum Geologists Search and Discovery Article 90063.

Lyu, C., Zhong, L., Ning, Z., Wang, Q., Cole, D.R., 2021. Characterization of nanoscale pores in tight gas sandstones using complex techniques: a case study of a Linxing tight gas sandstone reservoir. Geofluids 2021, . Available from: https://doi.org/10.1155/2021/7670556Article ID 7670556, 10 pages.

Mousavi, M.A., Bryant, S.L., 2012. Connectivity of pore space as a control on two-phase flow properties of tight-gas sandstones. Transport in Porous Media 94, 537–554.

Munawar, M.J., Lin, C., Cnudde, V., Bultreys, T., Dong, C., Zhang, X., et al., 2018. Petrographic characterization to build an accurate rock model using micro-CT: case study on low-permeable to tight turbidite sandstone from Eocene Shahejie Formation. Micron (Oxford, England: 1993) 109, 22–33.

Naik, G.C., 2003. Tight gas reservoirs—an unconventional natural energy source for the future. Accessadoem 1 (07), 2008Retrieved from. Available from: http://www.sublette-se.org/files/tight_gas.pdf.

Nelson, P.H., 2009. Pore-throat sizes in sandstones, tight sandstones, and shales. AAPG Bulletin 93 (3), 329–340.

Pittman, E.D., 1992. Relationship of porosity and permeability to various parameters derived from mercury injection-capillary pressure curves for sandstone. AAPG Bulletin 76 (2), 191–198.

Prodanovic, M., Bryant, S.L., Davis, J.S., 2013. Numerical simulation of diagenetic alteration and its effect on residual gas in tight gas sandstones. Transport in Porous Media 96, 39–62. Available from: https://doi.org/10.1007/s11242-012-0072-3.

Purcell, W.R., 1949. Capillary pressures-their measurement using mercury and the calculation of permeability therefrom, American Institute of Mechanical Engineers, Petroleum Transactions, February, 39–48. (SPE-949039-G).

Rahman, M.K., Suarez, Y.A., Chen, Z., Rahman, S.S., 2007. Unsuccessful hydraulic fracturing cases in Australia: investigation into causes of failures and their remedies. Journal of Petroleum Science and Engineering 57, 70–81.

Ren, D., Zhou, D., Liu, D., Dong, F., Shuwei Ma, S., Huang, H., 2019. Formation mechanism of the upper Triassic Yanchang formation tight sandstone reservoir in Ordos Basin —take Chang 6 reservoir in Jiyuan oil field as an example. Journal of Petroleum Science and Engineering 178, 497–505.

Rezaee, M.R., Jafari, A., Kazemzadeh, E., 2006. Relationships between permeability, porosity and pore throat size in carbonate rocks using regression analysis and neural networks. Journal of Geophysics and Engineering 3, 370–376.

Rezaee, R., Saeedi, A., Clennell, B., 2012. Tight gas sands permeability estimation from mercury injection capillary pressure and nuclear magnetic resonance data. Journal of Petroleum Science and Engineering 88–89, 92–99.

Rushing, J.A., Newsham, K.E., Blasingame, T.A., 2008. Rock typing: keys to understanding productivity in tight gas sands. In: SPE Unconventional Reservoirs Conference, Society of Petroleum Engineers.

Shanley, K.W., Cluff, R.M., Robinson, J.W., 2004. Factors controlling prolific gas production from low-permeability sandstone reservoirs: implications for resource assessment, prospect development, and risk analysis. AAPG Bulletin 88, 1083–1121.

Soeder, D.J., Chowdiah, P., 1990. Pore geometry in high-and low-permeability sandstones, Travis Peak Formation, east Texas: Society of Petroleum Engineers Formation. Evaluation 5 (4), 421–430.

Su, Y., Zha, M., Jiang, L., Ding, X., Qu, J., Jin, J., et al., 2022. Pore structure and fluid distribution of tight sandstone by the combined use of SEM, MICP and X-ray micro-CT. Journal of Petroleum Science and Engineering 208, 109241.

Sujuan, Y., Zhengxiang, L., Rong, L., 2011. Petrophysical and capillary pressure properties of the upper Triassic Xujiahe Formation tight gas sandstones in western Sichuan, China. Petroleum Science 8, 34–42.

Swanson, B.F., 1981. A simple correlation between permeability and mercury capillary pressures. Journal of Petroleum Technology 2488–2504.

Thomas, R.D., Ward, D.C., 1972. Effect of overburden pressure and water saturation on gas permeability of tight sandstone cores. Journal of Petroleum Technology 25 (2), 120–124.

Van Brakel, J., Modry, S., Svata, M., 1981. Mercury porosimetry: state of the art. Powder Technology 29, 1–12.

Walls, J.D., Nur, A.M., Bourbie, T., 1982. Effects of pressure and partial water saturation on gas permeability in tight sands: experimental results: paper SPE 9378-PA. Journal of Petroleum Technology 34 (4), 930–936. Available from: https://doi.org/10.2118/9378-PA.

Wang, H., Rezaee, R., Saeedi, A., 2015. Evaporation process and pore size distribution in tight sandstones: a study using NMR and MICP. Procedia Earth and Planetary Science 15, 767–773.

Wang, R., Shi, W., Xie, X., Zhang, W., Qin, S., Liu, K., et al., 2020. Clay mineral content, type, and their effects on pore throat structure and reservoir properties: insight from the Permian tight sandstones in the Hangjinqi area, north Ordos Basin, China. Marine and Petroleum Geology 115, 104281.

Washburn, E.W., 1921. The dynamics of capillary flow. Physical Review 17 (3), 273.

Wei, K.K., Morrow, N.R., Brower, K.R., 1986. Effect of fluid, confining pressure, and temperature on absolute permeabilities of low permeability sandstones: paper SPE 13093-PA. SPE Formation Evaluation 1 (4), 413–423. Available from: https://doi.org/10.2118/13093-PA.

Wu, H., Ji, Y.L., Zhang, C.L., Chen, S., 2017. Insight into the pore structure of tight gas sandstones: a case study in the Ordos Basin, NW China. Energy & Fuels: An American Chemical Society Journal 31 (12), 13159–13178.

Wu, Y., Tahmasebi, P., Lin, C., Zahid, M.A., Dong, C., Golab, A.N., et al., 2019. A comprehensive study on geometric, topological and fractal characterizations of pore systems in low-permeability reservoirs based on SEM, MICP, NMR, and X-ray CT experiments. Marine and Petroleum Geology 103, 12–28.

Wyllie, M.R., Gregory, A.R., Gardner, G.H.F., 1956. Elastic wave velocities in heterogeneous and porous media. Geophysics 21 (1), 41–70.

Xiao, Q., Yang, Z., Wang, Z., Qi, Z., Wang, X., Xiong, S., 2020. A full-scale characterization method and application for pore-throat radius distribution in tight oil reservoirs. Journal of Petroleum Science and Engineering 187, 106857.

Yang, P., Zhang, L., Liu, K., Cao, B., Gao, J., Qiu, G., 2021. Diagenetic history and reservoir evolution of tight sandstones in the second member of the Upper Triassic Xujiahe Formation, western Sichuan Basin, China. Journal of Petroleum Science and Engineering 201, 108451.
Zhang, L., Guoping Bai, G., Luo, X., Ma, X., Chen, M., Wu, M., Yang, W., 2009. Diagenetic history of tight sandstones and gas entrapment in the Yulin Gas Field in the central area of the Ordos Basin, China. Marine and Petroleum Geology 26, 974–989.
Zhao, X., Yang, Z., Lin, W., Xiong, S., Luo, Y., 2019. Study on pore structures of tight sandstone reservoirs based on nitrogen adsorption, high-pressure mercury intrusion, and rate-controlled mercury intrusion. Journal of Energy Resources Technology 141, 112903. 1.
Ziarani, A.S., Aguilera, R., 2012. Pore-throat radius and tortuosity estimation from formation resistivity data for tight-gas sandstone reservoirs. Journal of Applied Geophysics 83, 65–73.

Further reading

Bjgrlykke, K., Ramm, M., Saigal, G.C., 1989. Sandstone diagenesis and porosity modification during basin evolution. Geoiogische Rundschau 78, 243–268.
Castle, J.W., Byrnes, A.P., 2005. Petrophysics of lower silurian sandstones and integration with the tectonic-stratigraphic framework, Appalachian basin, United States. AAPG Bulletin 89 (1), 41–60.
Christopher, R., Clarkson, C.R., Jensen, J.L., Pedersen, P.K., Freeman, M., 2012. Innovative methods for flow-unit and pore-structure analyses in a tight siltstone and shale gas reservoir. AAPG Bulletin 96 (2), 355–374.
Liu, M., Xiong, C., 2021. Diagenesis and reservoir quality of deep-lacustrine sandy-debris-flow tight sandstones in upper Triassic Yanchang formation, Ordos Basin, China: implications for reservoir heterogeneity and hydrocarbon accumulation, Journal of Petroleum Science and Engineering, 202. p. 108548.
Martin, A.J., Solomon, S.T., Hartmann, D.J., 1997. Characterization of petrophysical flow units in carbonate reservoirs. AAPG Bulletin 81 (5), 734–759.
Stroker, T.M., Harris, N.B., Elliott, W.C., Wampler, J.M., 2013. Diagenesis of a tight gas sand reservoir: upper Cretaceous Mesaverde group, Piceance Basin, Colorado. Marine and. Petroleum Geology 40, 48–68.
Wu, H., Zhang, C., Ji, Y., Liu, R., Cao, S., Chen, S., et al., 2018. Pore throat characteristics of tight sandstones of Yanchang formation in eastern Gansu, Ordos Basin. Petroleum Research 3, 33–43.

CHAPTER 5

Porosity and permeability evolution

5.1 Introduction

Sandstones like other sedimentary rocks show a unique history of their formation and evolution. Three main factors controlling reservoir quality in sandstones are (1) depositional porosity and permeability under the effect of grain size and morphology, sorting, and sand to matrix ratio. (2) Compaction with regard to its degree and (3) pore-filling cement based on its amount and type (Worden and Morad, 2003). Similarly, tightness and low permeability in tight gas sandstone reservoirs can be attributed to depositional factors, mineralogy, and/or postdepositional processes during diagenesis. Depositional factors are related to sedimentary texture and sedimentary environment conditions (e.g., muddy sandstones in low-energy environments). Sandstones with low textural maturity (rich in matrix) such as graywackes are good examples in which tightness is mostly related to the rock matrix and depositional environment. Moreover, tight sandstone reservoirs, under the effect of postdepositional processes depict a modified form of original depositional facies and mineralogy. This is associated with changes in the initial reservoir properties of the rocks at the deposition time. In fact, the alteration of the pore system by diagenetic processes and their effects on porosity and permeability are major issues in tight gas reservoirs (Amann-Hildenbrand et al., 2012).

5.2 Depositional controls

Depositional factors include all sedimentary conditions related to the provenance, tectonic setting, sedimentary processes, and environments that control sediment composition, texture, and structure. As with the conventional reservoirs, sedimentary environments of tight sandstone reservoirs vary from the fluvial system, shoreline, and shallow marine (e.g., delta deposits) to deep marine (e.g., submarine fan and turbidite deposits). This means sandy deposits from the different depositional environments have the potential to tighten under conditions imposed by depositional/postdepositional factors. Table 5.1 list some tight sandstone reservoirs related to sedimentary

Reservoir Characterization of Tight Gas Sandstones
DOI: https://doi.org/10.1016/B978-0-323-90180-2.00002-5

Table 5.1 Some examples of a sedimentary environment of tight reservoir sandstones in different basins of the world.

Basin/field	Formation	Age	Sedimentary environment	References	Depth of reservoir (m)
Bohai Bay Basin, North China	Shihezi Formation	Permian	Fluvial to deltaic	Lv et al. (2011)	3100–3900
East Irish Sea Basin, United Kingdom	Collyhurst Sandstone Formation	Lower Permian	Aeolian and fluvial	Oluwadebi et al. (2018)	~3420
Perth Basin (Whicher Range field), Western Australia	Willespie Formation	Late Permian	Fluvial-lacustrine	Kadkhodaie-Ilkhchi et al. (2013)	~4000
Ordos Basin, China	Chang 7 Unit Yanchang Formation	Upper Triassic	Lacustrine basin	Guo et al. (2020)	~2323
Tarim Basin (Dibei gas field), Northwestern China	Ahe Formation	Lower Jurassic	shallow lacustrine-swamp	Pang et al. (2019)	–
East Texas Basin	Cotton Valley Sandstone	Upper Jurassic	Shallow Maine and fluvial-deltaic	Wescott (1983)	–
Vøring Basin of the Norwegian Sea	Nise Formation	Upper Santonian to Middle Campanian	Deep-sea fan system	Knaust (2009)	–
Greater Green River Basin (Wamsutter field)	Almond Formation	Late Cretaceous	Fluvial and coastal-plain to shallow marine	Tobin et al. (2010)	3048
Mediterranean Sea, Egypt	Tineh Formation	Oligocene	Deep marine	Abuamarah et al. (2019)	~6000

environments of different basins in the world. There are many cases in the literature that indicate the importance of depositional factors in controlling the reservoir properties of tight sandstone reservoirs. For example, by investigation of tight gas sandstones of the Khazzan Barik reservoir in northern Oman, and based on petrographic evidence, Millson et al. (2008) showed that despite a prolonged and complex burial history, primary depositional texture (grain size, sorting, and clay abundance) does continue to influence the reservoir quality. In addition, they indicated that sandstone's provenance and grain type directly impact the extent of secondary porosity networks, as feldspathic-rich intervals, seem to have better overall reservoir quality because of grain leaching and generation of a limited secondary porosity network. These pore networks may play an important function in controlling productivity. Furthermore, in deep marine and tight quartz and subfeldspathic wacke sandstones of the Mediterranean Sea in Egypt, filling the pore spaces with detrital clays is known as the most important reservoir quality-reducing factor (Abuamarah et al., 2019). Also, on a regional scale, provenance, tectonic setting, and transportation factors control the basin development, burial history, and textural and mineralogical maturity of sandstones. For example, in fault-controlled basins (where the clastic basins are adjacent to the provenance) such as the Moghan Basin in northwest Iran and the Nanxiang Basin of China (Zeng et al., 2010), where the sedimentation rate is high and thickness is large, sandstones are mainly characterized by poor textural and compositional maturity. These sandstones are prone to intense diagenesis (high compaction and cementation), and the development of tight and low permeability reservoirs. In contrast, in other basins such as the Appalachian foreland basin, the sandstones in areas with the lowest relative rate of subsidence show the highest porosity and permeability (Castle and Byrnes, 2005).

5.3 Diagenetic controls

Porosity reduction of sandstones during burial and under the effect of compaction and cementation forms the main part of the diagenetic history of many reservoir rocks and also tight sandstone reservoirs. On the other hand, dissolution and alteration of chemically unstable and metastable components of sandstones (e.g., Ca-Na feldspars and volcanic rock fragments) have significant control on qualitative and quantitative variations of pore system properties of reservoir rocks.

Similarly, the components and mineralogy of sandstones play an essential role in the diagenetic history and the evolution and modification of their pore systems. In fact, reservoir rocks depending on their mineralogy, show different behavior against diagenetic processes. In general, sandstones with low mineralogical maturity (rich in feldspars and unstable rock fragments) are more susceptible to tightness via the modification of their pore systems by compaction and development of authigenic types of cement such as clay minerals. Examples of sandstones with low mineralogical maturity are the lithic-feldspathic arenites of the Triassic Yanchang Formation of the Ordos Basin in China (Wang et al., 2020a) and the Eocene to Miocene Ojagheshlagh and Zeivar Formations of the Moghan Basin in the northwest of Iran, where mechanical compaction and authigenic clays (illite and chlorite) are the most important diagenetic processes for decreasing reservoir quality. On the other hand, in sandstone with high textural and mineralogical maturity, the presence of cement may have a significant effect on porosity destruction and tightness. As in mature and tight quartz arenites sandstones of the Baltic Basin with Cambrian age, quartz cement is one of the main causes of porosity loss (Laškova, 1987; Lis-Śledziona, 2019). The triangle diagram Fig. 5–1 shows different groups of sandstones in relation to dominant diagenetic

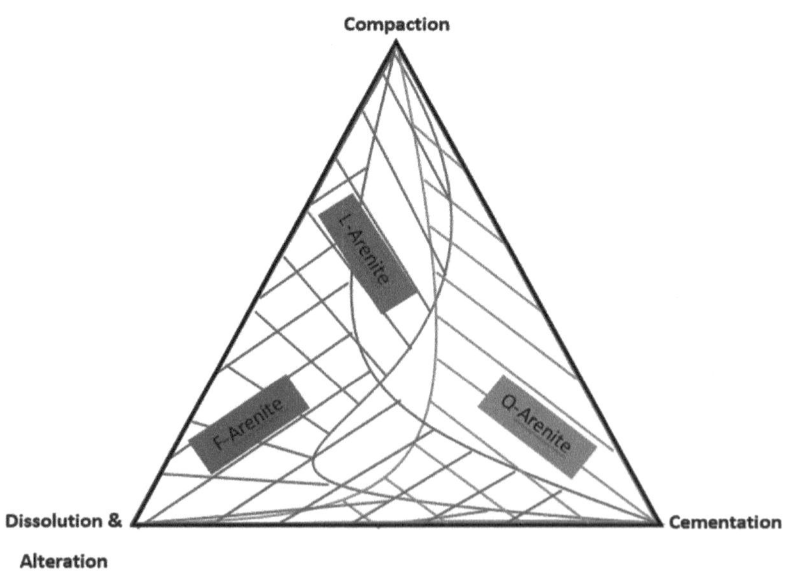

Figure 5–1 Graph showing the relationship between sandstone groups and the dominant diagenetic features. *Q*, Quartz; *F*, feldspar; *L*, lithic.

processes including compaction, cementation, dissolution, and alteration. As is seen, in quartz arenites that are rich in quartz grains, cementation especially by silica cement is a common diagenetic feature. Also, in feldspathic arenites, dissolution of feldspars and their alteration to clay minerals is more significant and in litharenites, the presence of flexible lithic components makes them more susceptible to compaction.

In the following section the impact of the main components of sandstones (minerals and rock fragments) on their diagenetic history and pore system evolution is described.

5.3.1 Quartz

Sandstones with a significant amount of quartz (e.g., quartz arenite) are resistant to compaction. However, other factors such as temperature and pressure, in addition to mineralogy, are involved in compaction degree, especially chemical compaction. The presence of overgrowth quartz around grains is a common diagenetic cement, mainly in quartz-rich sandstones (Fig. 5–2A). In these sandstones, quartz grains act as nucleation sites for the development of overgrowth cement. The abundance of quartz cement as a common cement, which destroys the intergranular porosity in deeply buried and quartz-rich sandstones, has been indicated in many sandstone reservoirs (e.g., Baron and Parnell, 2007; Mork and Moen, 2007; Kadkhodaie-Ilkhchi et al., 2013). This indicates the

Figure 5–2 SEM photomicrographs of (A) quartz overgrowth cement (Q) and (B) alteration of feldspar to kaolinite clay (K), and development of microporosity in tight sandstones of the late Permian Willespie Formation of the Perth Basin, Western Australia.

activation of various sources of silica for the development of this cement at burial depths and temperatures (e.g., Dutton and Loucks, 2010). Pressure solution, clay mineral transformation (e.g., smectite to illite), and alteration of detrital feldspars are some main burial sources involved in silica supply for cementation (e.g., Bjørkum et al., 1998; Gier et al., 2008; Islam, 2009; Metwally and Chesnokov, 2012). Chemical compaction that follows physical compaction with increasing burial depth, is characterized by pressure solution at grain contacts and stylolite as serrated surfaces lined by insoluble materials of the enclosing rock (Baron and Parnell, 2007; Tada and Siever, 1989).

5.3.2 Feldspars

Feldspar-rich sandstones such as arkose are more sensitive to chemical reactions by diagenetic fluids, which resulted in modification of their pore system through alteration, dissolution, and development of diagenetic products such as authigenic clay minerals and secondary pore types. Feldspar overgrowth cements (Fig. 5–3), with less importance and abundance than quartz overgrowths, are also observed in some feldspathic sandstones (e.g., Kim and Lee, 2004). Therefore their role as a pore-occluding cement is not significant. Due to the sensitivity of feldspars to alteration, they can act as a source for the development of authigenic clay cements as well as micropores in the pore systems of the sandstones (Fig. 5–2B). In the Lower Triassic tight sandstones from the Petrohan Terrigenous Group, NW Bulgaria, the presence of Illite has been attributed to the albitization of detrital K-feldspars (El-Ghali et al., 2009). Similarly, partly completely

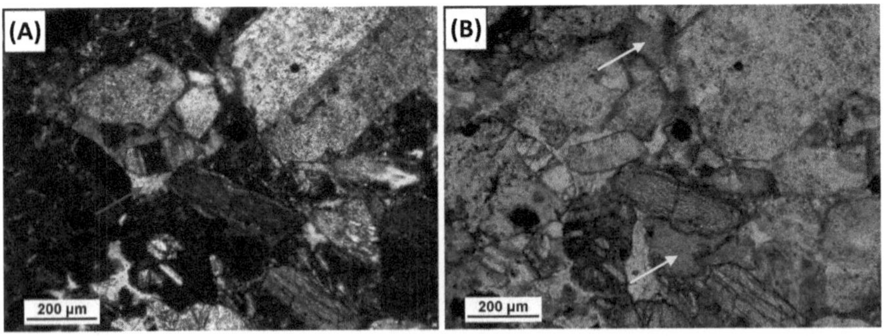

Figure 5–3 XPL (A) and PPL (B) photomicrographs of feldspar overgrowths (red arrow) cement in a feldspathic-lithic sandstone of the Zeivar Formation in the Moghan area, northwest Iran. Some secondary pores (blue-dyed) indicated by yellow arrows, derived from the dissolution of feldspars, are also observed.

leaching of feldspars during diagenesis, shows their contribution to the development of secondary pores in sandstone reservoirs (Figs. 5–3–5–5). In addition to provenance, the absence of feldspars in tight sandstones can be related to dissolution that is enhanced by some factors such as thermal events, basin subsidence, and organic acid derived from coal-bearing strata (Blake and Walter, 1996; Yuan et al., 2015).

According to the study results of Liu and Xiong (2021) on tight sandstones of the Upper Triassic Yanchang Formation of Ordos Basin in China, the creation of secondary pores from feldspar dissolution produce some byproducts occupying part of primary intergranular pores. They indicated that this process redistributes the pore types and covers the primary intergranular pores to dissolved pores and inter crystal micropores of authigenic clays, by which effective porosity and permeability slightly decrease. However, total porosity shows no clear trend with increasing feldspar dissolution porosity because the authigenic minerals derived from dissolution may precipitate elsewhere in the rock system (Molenar et al., 2007).

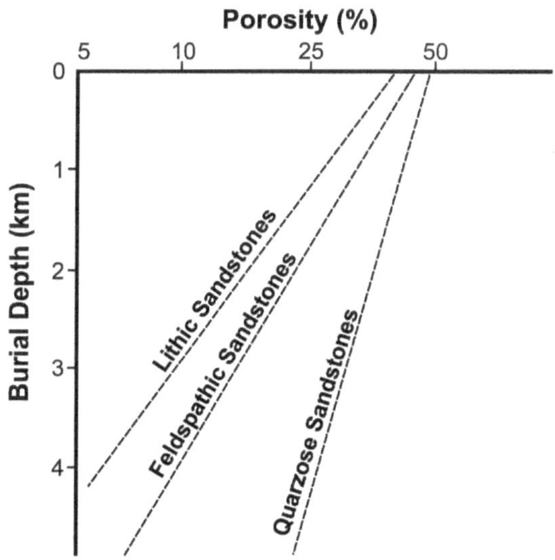

Figure 5–4 Graph showing the reduction rate of porosity as a function of burial depth (compaction) for sandstones of different compositions. The lower slope of the lithic sandstones line with respect to feldspathic and quartzose sandstones lines indicates the higher rate of porosity reduction with the burial depth of these sandstones (Dickinson, 1985). *Modified after Dickinson, W.R., 1985. Interpreting provenance relations from detrital modes of sandstones. In: Zuffa, GG. (Ed.), Provenance of Arenites. Reidel, Dordrecht, pp. 333–361.*

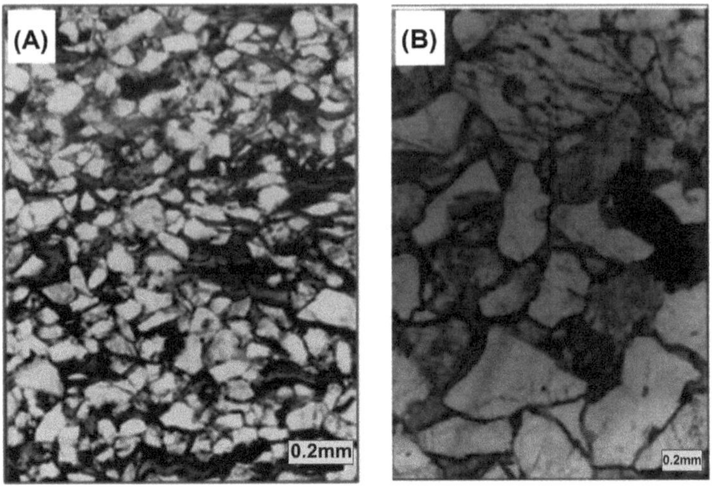

Figure 5–5 Photomicrographs of tight gas sandstones of the Whicher Range field in the Perth Basin showing different degrees of compaction on reservoir facies with different textures and mineralogy (Kadkhodaie-Ilkhchi et al., 2019). Fine-grained facies rich in ductile components between quartz grains (A) are more prone to compaction, compared to medium and coarse-grained sandstone facies with a rigid framework.

5.3.3 Rock fragments

Sandstones such as litharenites that are rich in unstable and ductile rock fragments (e.g., mudstone clasts, volcanic debris, and phyllitic and schistose metamorphic grains), in comparison with quartzose and feldspathic sandstones, are more prone to be affected by compaction. These sandstones lose their porosity at a faster rate during burial compared to quartzose and feldspathic sandstones (Gluyas, and Cade, 1997; Fig. 5–4). For example, in tight gas sandstones of the Whicher Range field in the Perth Basin of Western Australia (Kadkhodaie-Ilkhchi et al., 2019), and also in the Sulige gas field in the Central Ordos Basin of China (He et al., 2006), a part of the primary porosity between grains in coarse-grained facies, that forms the main reservoir part of the field, has been remained. This is attributed to the presence of coarse quartz grains and rock fragments that are resistant to compaction. In contrast, fine to medium-grained sandstones, which are also rich in ductile grains, have endured a high degree of compaction and thus significant reduction of primary porosity and permeability. In photomicrographs of Fig. 5–5 and core poroperm plot of Fig. 5–6, the effect of mineralogy and texture (grain size) on compaction

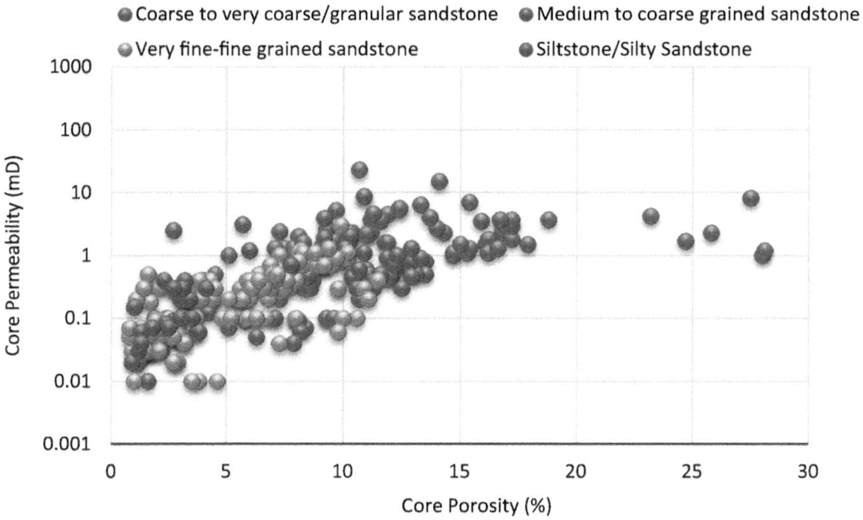

Figure 5–6 Core poroperm cross plot for tight gas sandstones of the Whicher Range field in the Perth Basin. According to the plot, coarse to medium-grained facies show higher values of porosity and permeability compared to fine and silty facies.

degree and reservoir quality is illustrated for the tight reservoir sandstones of the Whicher Range gas field. This is also observed in the tight litharenite sandstones of the Zeivar Formation in the Moghan area of northwestern Iran, where the effect of compaction due to the presence of volcanic rock fragments is significant and associated with a compacted fabric (Fig. 5–7A and B). Furthermore, the study results of Tobin et al. (2010) on lithic-rich tight sandstones of the Late Cretaceous Almond Formation in the Wamsutter field of Wyoming, indicated that compaction accounts for 54% (on average) of the total intergranular pore volume loss during burial. In this formation, the effect of compaction in the upper parts is more significant than in the middle and lower parts, which has been attributed to upwardly increasing amounts of lithic grains. Moreover, Li et al. (2021) mentioned deep burial and high content of rock fragments as the main causes for strong compaction and porosity reduction in tight sandstones of the Lower Silurian Kepingtage Formation of Tarim Basin in China. Generally, in ductile lithic-rich sandstones due to the dominant effect of compaction, diagenetic pore-filling cements are absent or limited to some diagenetic clays, as reported for a part of the Jurassic tight sandstone reservoirs in the central Junggar Basin of China (Hong et al., 2020).

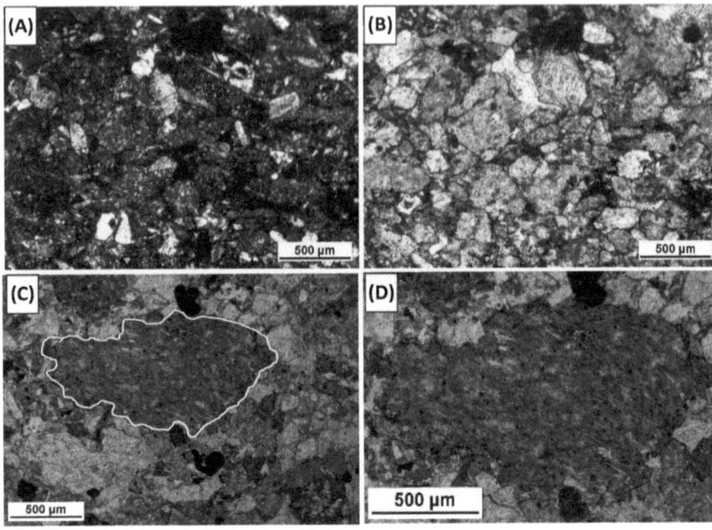

Figure 5–7 Photomicrographs of tight feldspathic-lithic sandstones of Moghan area in the northwest of Iran. (A) Volcanic litharenite with a compacted fabric in which significant impact of physical compaction has prevented the development of cements in later stages of diagenesis. (B) Photomicrograph A in PPL. (C) Development of secondary pores (blue-dyed) within the volcanic lithic (marked by the yellow boundary). (D) High resolution of volcanic lithic in photomicrograph (C). (For interpretation of the references to color in this figure legend, the reader is referred to the web version of this article.)

In addition to compaction, rock fragments based on their composition, are involved in the development of some diagenetic cements (e.g., clays and carbonates) and secondary pore types. For example, carbonate lithics and fragments (e.g., mud clasts) could be considered a source for the development of carbonate cements in sandstones during burial diagenesis (e.g., Morad, 1998; South and Talbot, 2000; Gier et al., 2008; Bril and Swennen, 2008). Moreover, some lithics such as volcanic types are sensitive to alteration by diagenetic fluids, by which authigenic clays are formed and microporosity and intragranular pores are added to the pore system of the sandstone reservoirs (Fig. 5–7C and D).

5.3.4 Micas

Micas are not frequent in sandstones, but their frequency is mainly dependent on the sediment source (provenance) and transportation factors. A high percent of micas in sandstones indicates the presence of mica-rich rocks (e.g., metamorphic rocks) in the source area and also short

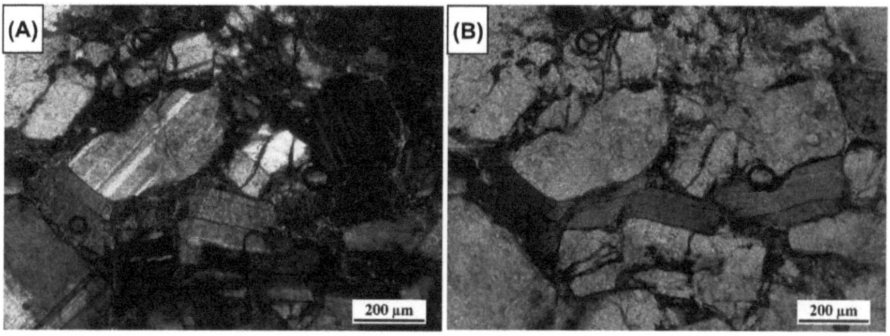

Figure 5–8 Photomicrographs (A) (XPL) and (B) (PPL) show a deformed mica in a tight and compacted feldspathic sandstone of the Zeivar Formation in the Moghan area, northwest Iran. The presence of soft and ductile mica in these facies has facilitated the effect of physical compaction.

transportation. In long transportations, micas due to their soft nature and high chemical reactivity, are removed from the sediment and are altered to very fine-grained minerals called Sericite. Similarly, the flexible structure makes micas more susceptible to physical compaction (Fig. 5–8), and thus the effect of compaction is enhanced by the occurrence of micas (El-Ghali et al., 2009). Alteration of micas to clay minerals is also a common diagenetic feature in mica-rich sandstones.

5.3.5 Clay minerals

Authigenic clay minerals are the most common diagenetic components of sandstones affecting their pore system properties. The importance and role of authigenic clays on reservoir properties of sandstones have been mentioned in the literature (e.g., Pittman, 1979; Bjørlykke, 1998; Khidir and Catuneanu, 2010; Wang et al., 2020b; Hong et al., 2020). Generally, the development of authigenic clays in primary intergranular pores reduces their effective porosity (Fig. 5–9). Therefore the study of authigenic clays is essential for reservoir characterization of tight sandstone reservoirs. Clay minerals import their effect on the pore system and tightness of sandstones through occluding pore spaces and pore throats, adding microporosity, increasing irreducible water saturation, and formation damage.

5.3.6 Occluding pore spaces and pore throats by clay minerals

Distribution of authigenic clays as grain replacement, pore-filling, pore-lining (grain coating), and pore-bridging determines their role and effect

Figure 5–9 Photomicrographs (A) (XPL) and (B) (PPL) showing the distribution of authigenic clay (chlorite) between grains in a litharenite sandstone of the Zeivar Formation in the Moghan area, northwest Iran.

in controlling reservoir properties. Although the presence of some grain-coating clays such as chlorite prevents the development of quartz overgrowth cement (e.g., Bloch et al., 2002; Taylor et al., 2010; Bjørlykke and Jahren, 2012; Oluwadebi et al., 2018; Hong et al., 2020), which results in preservation of primary porosity, their role in reducing effective porosity is more significant. The study results of Wang et al. (2020b) on the Permian tight sandstones in the Hangjinqi area of north Ordos Basin of China indicated the negative impact of clay mineral content (especially illite) on pore-throat and pore size distribution and porosity and permeability. They also mentioned that occluding and blocking the narrow throats by clay minerals is more significant than that in large pore types of these sandstones.

5.3.7 The effect of clay minerals on water saturation

Authigenic clays in addition to occluding pore spaces and pore throats, have a significant contribution to adding microporosity in the pore system of reservoir rocks, thus increasing water saturation. Micropores due to their complex structure and shape and also extremely small diameter (generally less than 2 μm) that increase the capillary displacement pressure, have little contribution to fluid flow in the reservoir (Guo et al., 2020). Moreover, clay minerals have a significant effect on increasing the internal specific surface area of the rock, by which the capacity for binding the fluid is increased (Wang et al., 2019). The role of clay minerals as an important agent in increasing immobile water saturation and the origin of low resistivity in some gas-bearing tight sandstone reservoirs has been mentioned in the literature (e.g., Cerepi et al., 2002; Jiang et al., 2021; Zeng et al., 2010).

5.3.8 The role of clay minerals in the formation of damage

Tight gas sandstone reservoirs suffer from formation damage or permeability damage that encounter the rock with permeability and production problems. In fact, under the effect of some factors such as the presence of authigenic clays, microfractures, small pore throats, and strong water wetting, tight reservoir sandstones are more prone to be damaged by water blocking and stress sensitivity at water-based underbalanced drilling (Feng et al., 2009). The main part of formation damage that is derived from drilling fluids or hydraulic fracturing fluids, is related to the presence of clay minerals in the reservoir. Water-sensitivity damage, water-block damage, and solid-phase damage are three main types of this damage. Clay minerals have an essential role in formation damage through expansion and providing dispersed particles and forming composite inclusions with fracturing fluid residues and high molecular polymers to block fractures and pores (Tang et al., 2021).

5.4 Paragenetic sequence

Paragenetic sequence describes the diagenetic history of sandstone reservoirs in terms of the various stages of diagenesis and related products developed in the reservoir. Eodiagenesis, mesodiagenesis, and telodiagenesis are three main stages of diagenesis, affecting the sediments after deposition, respectively. These stages are defined based on the physicochemical conditions (e.g., temperature, pressure, pore-water chemistry) that the sediments after deposition are experienced. Eodiagenesis includes all diagenetic processes just after deposition and during the shallow burial depths (about 2 km depth and temperature <70°C). Mesodiagenesis includes burial depths more than 2 km to the earliest stages of low-grade metamorphism and equivalent temperatures of about 200°C–250°C (Morad et al., 2000). Telodiagenesis is related to basin inversion and defines the diagenetic history of the rock during uplift and/or exposure due to tectonic activities.

Grain rotation and rearrangement and also deformation of ductile grains under the mechanical compaction are the main features of eodiagenesis, by which most of the initial porosity is reduced and a compacted fabric is created. Development of early diagenetic cements around and between grains after mechanical compaction may also occur during this stage. With increasing the burial depth and during mesodiagenesis, chemical compaction features at grain contacts are observed as solution seams and stylolites. Some burial cements such as carbonate cements and/or silica as overgrowth cement derived from pressure solution and mineral transformation and alteration of some components such as feldspars and rock fragments to authigenic clays are developed in this stage. Furthermore, secondary pores are formed by the dissolution of some unstable grains and early diagenetic cements. Compared to the conventional reservoirs, although tight gas sandstone reservoirs may show similar diagenetic sequences, the extent and intensity of the diagenetic processes that cause quantitative and qualitative changes in their pore systems are more significant. As the development of a compacted and cemented fabric with a modified pore system consists of limited remnants of initial intergranular pores, secondary dissolved pores with poor connection and micropores are the end product of these processes.

5.5 Burial history

The tectonic setting of sandstones has the main control on their burial history. In fact, variations in compaction degree are controlled by subsidence rate and burial history (Higgs et al., 2010). Knowledge of burial history provides information on gas generation and migration and primary destructive diagenetic agents such as compaction. In some tight gas sandstone reservoirs, a significant percentage of primary porosity and permeability are lost mainly due to compaction under the great burial depth. For example, in tight gas sandstones of the Perth Basin, physical and chemical compaction during the burial has had a significant effect on porosity loss. Furthermore, the high burial depths provide the conditions for the development of burial cements such as quartz overgrowths. As in the Permo-Triassic sandstones of Wagina Formation in the Northern Perth Basin of Western Australia, which has experienced burial depths over 4 km, extensive quartz cementation has strongly occluded pore spaces and pore throats (Laker, 2000). In assessing the main factor reducing the reservoir quality of tight sandstones in the Upper Triassic Yanchang Formation of Ordos Basin in China using a quantitative chart of Houseknecht (1987), Liu and Xiong (2021), found a positive relationship between burial depth and degree of compaction. Therefore they concluded that compaction has a more important role in reducing the intergranular porosity compared to cementation in this basin. Moreover, similar results have been mentioned for the tight sandstones of the Upper Triassic Xujiahe Formation, with a maximum burial depth of approximately 6000 m, in the western Sichuan Basin of China by Yang et al. (2021). Also, Oluwadebi et al. (2018) in their studies on the Lower Permian tight gas Collyhurst Sandstone Formation in the East Irish Sea Basin of northern England showed that the contribution of compaction in destroying the original porosity in two wells of the basin is 48.5% and 66.1% such that cementation accounts for 32.8% and 20.2% of the porosity destruction in these wells. The study results of Ivy (2018) on the tight Upper Carboniferous fluvial sandstones in the Lower Saxony Basin (NW Germany) indicated that the differences in burial histories in two fields of the basin have led to a reduction in the formation potential of authigenic quartz cement and thus a higher matrix porosity is observed in the field with stronger uplift during basin inversion. Moreover, according to Shanley et al. (2007), in low permeability sandstone reservoirs from Rocky Mountain basins of Western America, hydrocarbon charge and migration have occurred before maximum burial depth, when many reservoirs were not tight. However, during burial, with decreasing pore volume, water saturations and gas column heights were increased and reservoirs changed to tight. During uplift, some traps spilled, and gas column adjusted to uplift and increased and many capillary systems departed from primary drainage to imbibition. Therefore, based on the results of these studies, it can be stated that in most tight sandstone reservoirs, deep burial depths have provided the conditions for effective operation or intensification of diagenetic processes.

5.6 Reservoir heterogeneity

Heterogeneity in sandstone reservoirs that is mainly controlled by depositional, diagenetic and structural (fractures and faults) factors, is defined in different scales including microscopic (μm), macroscopic (in meters), megascopic (10–100 m), and gigascopic (> 300 m) scales (Morad et al., 2010). Heterogeneity of tight sandstones can be considered from different aspects, including lithology, pore structure, petrophysical and reservoir properties. These parameters are closely related to each other, as heterogeneity in lithological parameters (texture and composition) controls heterogeneity of pore structure, and at the same time, both of them have significant control on heterogeneity related to petrophysical and reservoir properties. The heterolithic nature observed in some tight sandstone reservoirs as repetitive intervals consisting of different lithology such as sandstone, siltstone, and shale as well as vertical changes in depositional patterns as fining- or coarsening-upward sequences is responsible for part of the reservoir heterogeneity related to depositional environment conditions. Another case of this heterogeneity is the local presence of high-quality intervals within tight sandstone reservoirs, as mentioned in the literature review of these reservoirs (e.g., Shanley et al., 2004; Zhang, 2008; Lai et al., 2015). Moreover, diagenesis has a significant control on reservoir heterogeneity of tight sandstones by the development of diagenetic materials and qualitative and quantitative changes in the pore system of reservoir rocks resulting in variation of porosity-permeability relationships. The importance of diagenetic processes on reservoir heterogeneity of tight sandstone reservoirs has been mentioned in different basins of the world. For example, Khidir and Catuneanu (2010) indicated the impact of calcite cementation on reservoir quality and heterogeneity of the late Maastrichtian-Early Paleocene Scollard Formation in the foredeep of the Western Canada Sedimentary Basin. Also, Kadkhodaie-Ilkhchi et al. (2019) illustrated the effect of diagenesis on controlling the internal reservoir heterogeneity by modification of pore system properties in tight gas sandstones of the Perth Basin in Western Australia. Furthermore, the role of diagenesis on heterogeneity has been mentioned in tight sandstone reservoirs of different basins in China such as the Sichuan Basin (Liu et al., 2018; Yang et al., 2021) and Ordos Basin (Li et al., 2021; Liu and Xiong, 2021). Therefore investigation of diagenetic facies can help for unraveling the diagenetic heterogeneity of tight sandstones. The final effect of diagenetic processes on reservoir rocks is recorded in pore network properties as a change in volume, type, and connection of pore spaces. Pore-scale heterogeneity has a significant effect on gas flow and hydrocarbon deliverability (Qiao et al., 2021). Wang et al. (2020b) consider the weak correlation between porosity and permeability of tight sandstones as evidence of the strong heterogeneity of these reservoirs. In fact, the extreme heterogeneity of tight gas sandstones due to the complex pore characteristics (i.e., size, shape, location, and connectivity of pores and throats), has encountered these reservoirs with exploration and development issues (Higgs et al., 2007; Huang et al., 2017). Variations in lithology (composition and texture), diagenetic features, and pore network properties of tight sandstones result in various rock types with different reservoir and petrophysical properties. Cao et al. (2021) indicate the importance of petrofacies analysis (texture and diagenesis) for a better understanding of the factors controlling reservoir quality and pore structures of tight sandstones. Therefore investigation of depositional facies and diagenetic features and pore structure analysis in the framework of reservoir rock typing provides an effective tool for unraveling the reservoir heterogeneity of tight gas sandstone reservoirs.

References

Abuamarah, B.A., Nabawy, B.S., Shehata, A.M., Kassem, O.M.K., Ghrefat, H., 2019. Integrated geological and petrophysical characterization of Oligocene deep marine unconventional poor to tight sandstone gas reservoir. Marine and Petroleum Geology 109, 868–885.

Amann-Hildenbrand, A., Ghanizadeh, A., Krooss, B.M., 2012. Transport properties of unconventional gas systems. Marine and Petroleum Geology 31 (1), 90–99.

Baron, M., Parnell, J., 2007. Relationships between stylolites and cementation in sandstone reservoirs: examples from the North Sea, UK and East Greenland. Sedimentary Geology 194 (1–2), 17–35.

Bjørkum, P.A., Oelkers, E.H., Nadeau, P.H., Walderhaug, O., Murphy, W.M., 1998. Porosity prediction in quartzose sandstones as a function of time, temperature, depth, stylolite frequency, and hydrocarbon saturation. AAPG Bulletin 82, 637–648.

Bjørlykke, K., 1998. Clay mineral diagenesis in sedimentary basins; a key to prediction of rock properties. Examples from the North Sea Basin. Clay Minerals 33, 15–34.

Bjørlykke, K., Jahren, J., 2012. Open or closed geochemical systems during diagenesis in sedimentary basins: constraints on mass transfer during diagenesis and the prediction of porosity in sandstone and carbonate reservoirs. AAPG (American Association of Petroleum Geologists) Bulletin. 96 (12), 2193–2214.

Blake, R.E., Walter, L.M., 1996. Effects of organic acids on the dissolution of orthoclaseat 80°C and pH6. Chemical Geology 132, 91–102.

Bloch, S., Lander, R.H., Bonnell, L., 2002. Anomalously high porosity and permeability in deeply buried sandstone reservoirs: origin and predictability. AAPG (American Association of Petroleum Geologists) Bulletin. 86 (2), 301–328.

Bril, K.V.D., Swennen, R., 2008. Sedimentological control on carbonate cementation in the Luxembourg sandstone formation. Geologica Belgica 12/1–2, 3–23.

Cao, B., Sun, W., Li, J., 2021. Reservoir petrofacies—a tool for characterization of reservoir quality and pore structures in a tight sandstone reservoir: a study from the sixth member of Upper Triassic Yanchang Formation, Ordos Basin, China. Journal of Petroleum Science and Engineering 199, 108294.

Castle, J.W., Byrnes, A.P., 2005. Petrophysics of lower Silurian sandstones and integration with the tectonic-stratigraphic framework, Appalachian basin, United States. AAPG Bulletin 89 (1), 41–60.

Cerepi, A., Durand, C., Brosse, E., 2002. Pore microgeometry analysis in low-resistivity sandstone reservoirs. Journal of Petroleum Science and Engineering 35 (3–4), 205–232.

Dickinson, W.R., 1985. Interpreting provenance relations from detrital modes of sandstones. In: Zuffa, G.G. (Ed.), Provenance of Arenites. Reidel, Dordrecht, pp. 333–361.

Dutton, S.P., Loucks, R.G., 2010. Diagenetic controls on evolution of porosity and permeability in Lower Tertiary Wilcox sandstones from shallow to ultradeep (200–6700 m) burial, Gulf of Mexico Basin, USA. Marine and Petroleum Geology 27, 69–81.

El-Ghali, M.A.K., Morad, S., Mansurbeg, H., Caja, M.A., Ajdanlijsky, G., Neil, O., et al., 2009. Distribution of diagenetic alterations within depositional facies and sequence stratigraphic framework of fluvial sandstones: Evidence from the Petrohan Terrigenous Group, Lower Triassic, NW Bulgaria. Petroleum Geology 26, 1212–1227.

Feng, Z., Hongming, T., Yingfeng, M., Gao, L., Xijin, X., 2009. Damage evaluation for water-based underbalanced drilling in low-permeability and tight sandstone gas reservoirs. Petroleum Exploration and Development 36 (1), 113–119.

Gier, S., Worden, R.H., Johns, W.D., Kurzweil, H., 2008. Diagenesis and rese rvoir quality of Miocene sandstones in the Vienna Basin, Austria. Marine and Petroleum Geology 25, 681–695.

Gluyas, J., Cade, C.A., 1997. Prediction of porosity in compacted sands. In: Kupecz, J.A., Gluyas, J., Bloch, S. (Eds.), Reservoir Quality Prediction in Sandstones and Carbonates, 69. AAPG Memoir, pp. 19–28.

Guo, R., Xie, Q., Qu, X., Chu, M., Li, S., Ma, D., et al., 2020. Fractal characteristics of pore-throat structure and permeability estimation of tight sandstone reservoirs: a case study of Chang 7 of the Upper Triassic Yanchang Formation in Longdong area, Ordos Basin, China. Journal of Petroleum Science and Engineering 184, 106555.

He, D., Jia A., Xu, T., 2006. Reservoir characterization of the giant sulige gas field, Ordos Basin, China. In: Poster Presentation at American Association of Petroleum Geologists Bulletin Annual Convention, Houston, Texas.

Higgs, K.E., et al., 2007. Diagenesis, porosity evolution, and petroleum emplacement in tight gas reservoirs, Taranaki Basin, New Zealand. Journal of Sedimentary Research 77, 1003–1025.

Higgs, K.E., Arnot, M.J., Browne, G.H., Kennedy, E.M., 2010. Reservoir potential of Late Cretaceous terrestrial to shallow marine sandstones, Taranaki Basin, New Zealand. Marine and Petroleum Geology 27, 1849–1871.

Hong, D., Cao, J., Wu, T., Dang, S., Hu, W., Yao, S., 2020. Authigenic clay minerals and calcite dissolution influence reservoir quality in tight sandstones: insights from the central Junggar Basin, NW China. Energy Geoscience 1, 8–19.

Houseknecht, D.W., 1987. Assessing the relative importance of compaction processes and cementation to reduction of porosity in sandstones. AAPG Bulletin 71, 633–642.

Huang, W., Shuangfang, L., Hersi, O.S., Wang, M., Deng, S., Lu, R., 2017. Reservoir spaces in tight sandstones: classification, fractal characters, and heterogeneity. Journal of Natural Gas Science and Engineering 46, 80–92.

Islam, M.A., 2009. Diagenesis and reservoir quality of Bhuban sandstones (Neogene), Titas Gas Field, Bengal Basin, Bangladesh. Journal of Asian Earth Sciences 35 (1), 89–100.

Ivy, B., 2018. Structural and Diagenetic Controls on Reservoir Quality in Tight Siliciclastic and Carbonate Rocks. KIT Scientific Publishing, p. 214.

Jiang, Y., Cui, J., Zhou, J., Fu, X., Cui, L., Fang, C., 2021. Analyzing the origin of low resistivity in gas-bearing tight sandstone reservoir. Geofluids 2021. Available from: https://doi.org/10.1155/2021/4341804Article ID 4341804.

Kadkhodaie-Ilkhchi, R., Kadkhodaie-Ilkhchi, A., Rezaee, R., 2019. Unraveling the reservoir heterogeneity of the tight gas sandstones using the porosity conditioned facies modeling in the Whicher Range field, Perth Basin, Western Australia. Journal of Petroleum Science and Engineering 176, 97–115.

Kadkhodaie-Ilkhchi, R., Rezaee, R., Moussavi-Harami, R., Kadkhodaie-Ilkhchi, A., 2013. Analysis of the reservoir electrofacies in the framework of hydraulic flow units in the Whicher Range Field, Perth Basin, Western Australia. Journal of Petroleum Science and Engineering 111, 106–120.

Khidir, A., Catuneanu, O., 2010. Reservoir characterization of Scollard-age fluvial sandstones, Alberta foredeep. Marine and Petroleum Geology 27, 2037–2050.

Kim, Y., Lee, Y.I., 2004. Diagenesis of shallow marine sandstones, the Lower Ordovician Dongjeom formation, Korea: response to relative sea-level changes, Journal of Asian Earth Sciences, 23. pp. 235–245.
Knaust, D., 2009. Characterization of a deep-sea fan system in the Norwegian Sea by means of ichnofabrics. Marine and Petroleum Geology 26, 1199–1211.
Lai, J., Wang, G., Ran, Y., Zhou, Z., 2015. Predictive distribution of high-quality reservoirs of tight gas sandstones by linking diagenesis to depositional facies: evidence from Xu-2 sandstones in the Penglai area of the central Sichuan Basin, China. Journal of Natural Gas Science and Engineering 23, 97–111.
Laker, N., 2000. Sedimentological and petrographic review of the Permo-Triassic sandstones of the Northern Perth Basin. Prepared for Australian Worldwide Exploration Limited, unpublished Report, pp. 150.
Laškova, LN., 1987. The Cambrian: oil fields of the Baltic region. Mokslas-Vilnius, 10–22.
Li, B., Zhang, H., Xia, Q., Peng, J., Guo, Q., 2021. Quantitative evaluation of tight sandstone reservoir based on diagenetic facies -a case of lower Silurian Kepingtage formation in Shuntuoguole low uplift, Tarim Basin, China. Frontiers in Earth Science 8, 597535.
Lis-Śledziona, A., 2019. Petrophysical rock typing and permeability prediction in tight sandstone reservoir. Acta Geophysica 67, 1895–1911.
Liu, Y., Hu, W., Cao, J., Wang, X., Tang, Q., Wu, H., et al., 2018. Diagenetic constraints on the heterogeneity of tight sandstone reservoirs: a case study on the upper Triassic Xujiahe formation in the Sichuan Basin, southwest China. Marine and Petroleum Geology 92, 650–669.
Liu, M., Xiong, C., 2021. Diagenesis and reservoir quality of deep-lacustrine sandy-debris-flow tight sandstones in Upper Triassic Yanchang Formation, Ordos Basin, China:implications for reservoir heterogeneity and hydrocarbon accumulation, Journal of Petroleum Science and Engineering, 202. p. 108548.
Lv, D., Li, Z., Chen, J., Liu, H., Guo, J., Shang, L., 2011. Characteristics of the permian coal-formed gas sandstone reservoirs in Bohai Bay Basin and the adjacent areas, North China. Journal of Petroleum Science and Engineering 78, 516–528.
Metwally, Y.M., Chesnokov, E.M., 2012. Clay mineral transformation as a major source for authigenic quartz in thermo-mature gas shale. Applied Clay Science 55, 138–150.
Millson, J.A., Quin, J.G., Idiz, E., Turner, P., Al-Harthy, A., 2008. The Khazzan gas accumulation, a giant combination trap in the Cambrian Barik Sandstone Member, Sultanate of Oman: Implications for Cambrian petroleum systems and reservoirs. AAPG Bulletin 92 (7), 885–917.
Molenar, N., Cyziene, J., Sliaupa, S., 2007. Quartz cementation mechanisms and porosity variation in Baltic Cambrian sandstones. Sedimentary Geology 195 (3–4), 135–159.
Morad, S., 1998. Carbonate cementation in sandstones: distribution patterns and geochemical evolution. In: Morad, S. (Ed.), Carbonate Cementation in Sandstones: Distribution Patterns And Geochemical Evolution, 26. International Association of Sedimentologists, Special Publication, pp. 1–26.
Morad, S., Al-Ramadan, K., Ketzer, J.M., De Ros, L.F., 2010. The impact of diagenesis on the heterogeneity of sandstone reservoirs: a review of the role of depositional facies and sequence stratigraphy. AAPG Bulletin 94 (8), 1267–1309.
Morad, S., Ketzer, J.M., De Ros, F., 2000. Spatial and temporal distribution of diagenetic alterations in siliciclastic rocks: implications for mass transfer in sedimentary basins, Sedimentology, 47. pp. 95–120.
Mork, M.B.E., Moen, K., 2007. Compaction microstructures in quartz grain and quartz cement in deeply buried reservoir sandstone using combined petrology and EBSD analysis. Journal of Structural Geology 29, 1843–1854.
Oluwadebi, A.G., Taylor, K.G., Dowey, P.J., 2018. Diagenetic controls on the reservoir quality of the tight gas Collyhurst Sandstone formation, lower Permian, East Irish Sea Basin, United Kingdom. Sedimentary Geology 371, 55–74.

Pang, X., Peng, J., Jiang, Z., Yang, H., Wang, P., Jiang, F., et al., 2019. Hydrocarbon accumulation processes and mechanisms in Lower Jurassic tight sandstone reservoirs in the Kuqa subbasin, Tarim Basin, northwest China: A case study of the Dibei tight gas field. AAPG Bulletin 103 (4), 769–796.

Pittman, E.F., 1979. Porosity, diagenesis and productive capability of sandstone reservoirs. Society of Economic Paleontologists and Mineralogists Special Publication 26, 159–173.

Qiao, J., Zeng, J., Cai, J., Jiang, S., An, T., Xiao, E., et al., 2021. Pore-scale heterogeneity of tight gas sandstone: origins and impacts. Journal of Natural Gas Science and Engineering 96, 104248.

Shanley, K.W., Cluff, R.M., Robinson, J.W., 2004. Factors controlling prolific gas production from low-permeability sandstone reservoirs: implications for resource assessment, prospect development, and risk analysis. AAPG (American Association of Petroleum Geologists) Bulletin. 88, 1083–1121.

Shanley, K.W., Cluff, R.M., Robinson, J.W., 2007. Prolific gas production from low permeability sandstone reservoirs – part ii: reconciling basin history, fluid saturations, gas shows, and capillary pressure. In: Presentation at Session, "Searching for Success," AAPG 2007 Annual Convention, Long Beach, California, April 1–4, 2007.

South, D.L., Talbot, M.R., 2000. The sequence stratigraphic framework of carbonate diagenesis within transgressive fan-delta deposits: Sant Llorenç del Munt fan-delta complex, SE Ebro Basin, NE Spain. Sedimentary Geology 138, 179–198.

Tada, R., Siever, R., 1989. Pressure solution during diagenesis. Annual Review of Earth and Planetary Sciences 17, 89–118.

Tang, H., Tang, H., He, J., Zhao, F., Zhang, L., Liao, J., et al., 2021. Damage mechanism of water-based fracturing fluid to tight sandstone gas reservoirs: Improvement of the evaluation measurement for properties of water-based fracturing fluid: SY/T 5107-2016. Natural Gas Industry B 8 (2), 163–172.

Taylor, T.R., Giles, M.R., Hathon, L.A., et al., 2010. Sandstone diagenesis and reservoir quality prediction: models, myths, and reality. AAPG (American Association of Petroleum Geologists) Bulletin 94 (8), 1093–1132.

Tobin, R.C., McClain, T., Lieber, R.B., Ozkan, A., Banfield, L.A., Marchand, A.M.E., et al., 2010. Reservoir quality modeling of tight-gas sands in Wamsutter field: integration of diagenesis, petroleum systems, and production data. AAPG Bulletin 94 (8), 1229–1266.

Wang, J., Wu, S., Li, Q., Xiao, S., 2020a. Controls of diagenetic alteration on the reservoir quality of tight sandstone reservoirs in the Triassic Yanchang Formation of the Ordos Basin, China. Journal of Asian Earth Sciences 200, 104472.

Wang, J., Wu, S., Li, Q., Zhang, J., Guo, Q., 2020b. Characterization of the pore-throat size of tight oil reservoirs and its control on reservoir physical properties: a case study of the Triassic tight sandstone of the sediment gravity flow in the Ordos Basin, China. Journal of Petroleum Science and Engineering 186, 106701.

Wang, M., Yang, Z., Shui, C., Yu, Z., Wang, Z., Cheng, Y., 2019. Diagenesis and its influence on reservoir quality and oil-water relative permeability: a case study in the Yanchang formation Chang8 tight sandstone oil reservoir, Ordos Basin, China. Open Geoscience 11, 37–47.

Wescott, W.A., 1983. Diagenesis of cotton valley sandstone (Upper Jurassic), East Texas: Implications for tight gas formation pay recognition. AAPG Bulletin 67 (6), 1002–1013.

Worden, R.H., Morad, S., 2003. Clay minerals in sandstones: controls on formation, distribution and evolution. International Association of Sedimentologists 34, 3–41.

Yang, P., Zhang, L., Liu, K., Cao, B., Gao, J., Qiu, G., 2021. Diagenetic history and reservoir evolution of tight sandstones in the second member of the upper Triassic Xujiahe formation, western Sichuan Basin, China. Journal of Petroleum Science and Engineering 201, 108451.

Yuan, G.H., Cao, Y.C., Gluyas, J., Li, X.Y., Xi, K.L., Wang, Y.Z., et al., 2015. Feldspar dissolution, authigenic clays, and quartz cements in open and closed sandstone geochemical systems during diagenesis: typical examples from two sags in Bohai Bay Basin, East China. AAPG Bulletin 99, 2121–2154.

Zeng, L., Jiang, J., Yang, Y., 2010. Fractures in the low porosity and ultra-low permeability glutenite reservoirs: a case study of the late Eocene Hetaoyuan formation in the Anpeng Oil field, Nanxiang Basin, China. Marine and Petroleum Geology 27, 1642–1650.

Zhang, S., 2008. Tight sandstone gas reservoirs: their origin and discussion. Oil Gas Geology 29, 1–11.

Further reading

Dutton, S.P., Clift, S.J., Hamilton, D.S., Hamlin, H.S., Hentz, T.F., Howard, W.E., et al., 1993. Major low-permeability sandstone gas reservoirs in the continental United States. In: Bureau of Economic Geology, Report of Investigations No. 211, The University of Texas at Austin, pp. 221.

John, A., Millson, J.A., Quin, J.G., Idiz, E., Turner, P., Al-Harthy, A., 2008. The Khazzan gas accumulation, a giant combination trap in the Cambrian Barik sandstone member, sultanate of Oman: implications for Cambrian petroleum systems and reservoirs. AAPG Bulletin 92 (7), 885–917.

Zhang, L., Bai, G., Luo, X., Ma, X., Chen, M., Wu, M., et al., 2009. Diagenetic history of tight sandstones and gas entrapment in the Yulin Gas Field in the central area of the Ordos Basin, China. Marine and Petroleum Geology 26, 974–989.

CHAPTER 6

The role of natural fractures in reservoir quality assessment of tight gas sandstones

6.1 Introduction

Natural fractures can occur in tight gas reservoirs. They play an important role in migration, entrapment, and production from tight gas sandstones. In tight gas sandstones, due to the very low matrix permeability, the drainage of the field is highly dependent on a well-connected, open fracture network. Fluid-flow rate and recovered resources are substantially impacted by the presence of the natural fracture. Fractures can occur parallel, oblique, or perpendicular to the bedding. Bedding fractures (also called interlayer fractures) aid in the horizontal migration of gas, while oblique and vertical fractures, also called tectonic fractures, contribute to the vertical migration (Fig. 6—1). Production from a tight gas sandstone heavily relies on the development of the fracture. Natural fractures propagate parallel to the direction of maximum horizontal stress, while their opening is parallel to the direction of minimum horizontal stress. Three generations of natural fractures occur in an anticlinal trap hosting tight gas sandstones, including hinge-perpendicular, hinge-oblique, and hinge-parallel (Price, 1966). Fractures perpendicular to the hinge line are prefolding, while hinge-oblique and hinge-parallel fractures are postfolding or fold-related fractures. Price's classification of fractures typical for asymmetric anticlines (after Price, 1966; Price & Cosgrove, 1990; Awdal et al., 2016) is shown in Fig. 6—2. Subsurface fracture studies are feasible through the interpretation of cores, image logs, and seismic data, while surface fracture studies are implemented through outcrop studies of the tight gas sandstones. Integration of outcrop studies with subsurface data provides fundamental inputs for fractures characterization and modeling.

6.2 Outcrops and understanding the subsurface natural fractures

Outcrop studies are essential for fracture characterization of tight gas sandstones. In outcrop studies, sampling sites are easily accessible, and frequent

Reservoir Characterization of Tight Gas Sandstones
DOI: https://doi.org/10.1016/B978-0-323-90180-2.00003-7

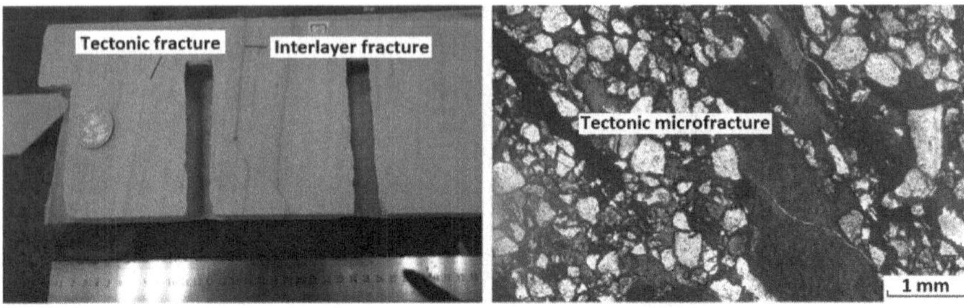

Figure 6–1 Tectonic fracture cut interlayer fractures in core observation from the Huagang tight gas sandstone, the Xihu Sag, East China Sea (left). Photomicrograph showing oblique tectonic fractures (right) (Zhou et al., 2020).

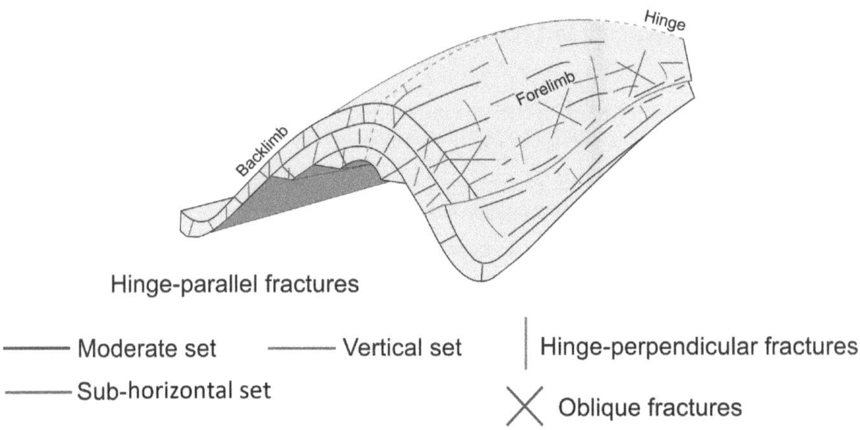

Figure 6–2 Observed fracture sets shown in Price's classification of fracture typical for asymmetric anticlines. *After Price, N., 1966, Fault and Joint Development in Brittle and Semi-Brittle Rock. Pergamon Press, Oxford, pp. 176.; Awdal, A., Healy, D., Alsop, G.I., 2016. Fracture patterns and petrophysical properties of carbonates undergoing regional folding: a case study from Kurdistan. N Iraq Journal of Marine and Petroleum Geology 71, 149–167.*

specimens and fracture measurements can be achieved (Corbett et al., 1987; Gutmanis et al., 2018). There is a good agreement between the fracture parameters such as orientation, intensity, spacing, and cement from outcrops and subsurface studies (e.g., cores, image logs, well logs, etc.). A realistic estimation of fracture length and azimuth is obtained from outcrop studies as it is not feasible to measure them from cores or image logs. However, some discrepancies exist between the surface and subsurface fracture studies due to the burial, uplift, erosion, and thermal effects causing differences in fracture growth and its mechanical properties

(Pollard and Aydin, 1988). A representation of steep and horizontal fractures from the outcrops and cores with an example from tight gas sandstones of the Xujiahe Formation, in the Sichuan Basin of China, is shown in Fig. 6–3.

6.3 Fractures characterization from borehole Image logs

Fractures are planes of separation with no displacement of rock blocks in geological formations. Borehole image logs [e.g., Formation Micro-Imager (FMI), STAR, UBI, and CAST] provide beneficial information for fracture characterization in tight gas sandstones. Image logs as advanced

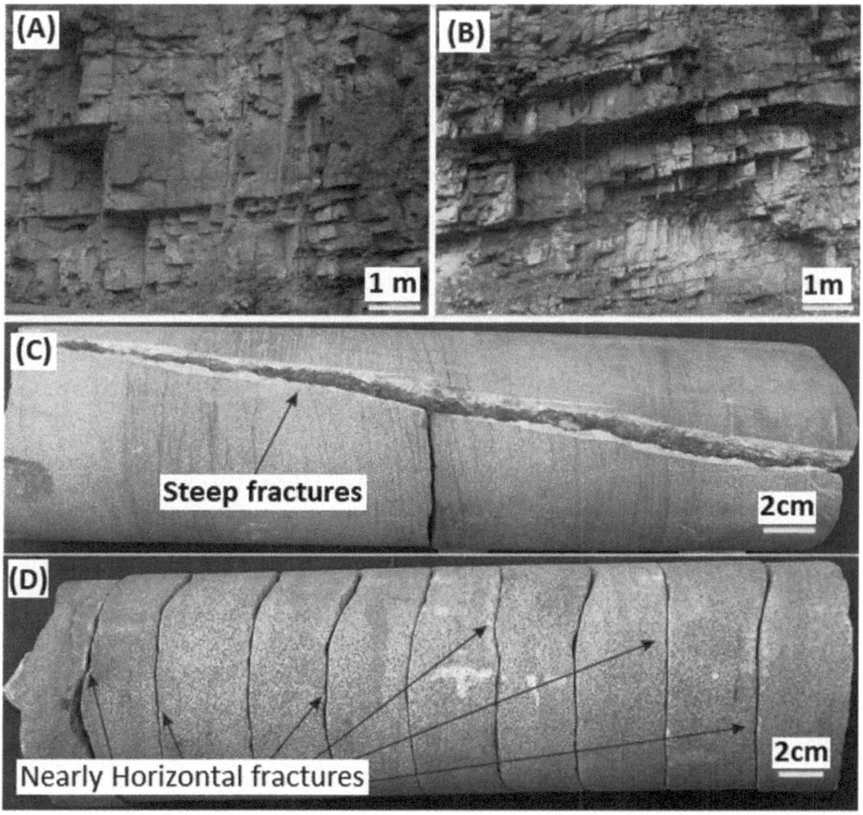

Figure 6–3 Representation of fractures in outcrops and cores. Steep fractures are seen in outcrop (A & B), nearly vertical fracture-filled with calcite cement in core (C), horizontal fractures in core (D), tight gas sandstones of the Xujiahe Formation, Sichuan Basin, China (Gong et al., 2019).

data acquisition tools provide a circumferential image of the physical properties of a borehole wall based on the electrical resistivity or acoustic impedance contrast. An example of Schlumberger's Fullbore FMI tool along with the resistivity electrodes laid on the pads and flaps is shown in Fig. 6–4. As is seen, the tool is equipped with four arms and four flaps connected to each arm. A total of 24 electrical resistivity buttons are laid on each pad or flap of the Schlumberger's FMI tool resulting in 192 resistivity measurements for each depth point.

Open fractures interpreted from the water-based image logs (e.g., FMI logs) or acoustic-based image logs (e.g., UBI) are seen as dark-colored sinusoids (Fig. 6–5). Conductive drilling fluids fill the fracture openings and cause a drop in electrical resistivity marked by black colors. In the case of acoustic image logs, fractured zones cause an increase in the acoustic transit time which is shown in black colors. In the oil-based image logs, the high electrical resistivity drilling fluid fills the fractures causing a resistive feature compared to the intact matrix that is shown in white colors. As shown in Figs. 6–6 and 6–7, fractures are seen as sine waves or sinusoids in the image logs. Projection of a fracture plane into an unrolled borehole cylinder results in sinusoids indicating fractures. Fractures filled with organic matter of dense material such as anhydrite and calcite appear resistive, while clay minerals or pyrite make cause conductive responses in image logs. To discriminate the pyrite-filled fractures from the mud filtrate-filled fractures that appear as conductive traces in image logs, the macroscopic and microscopic core and description data are required.

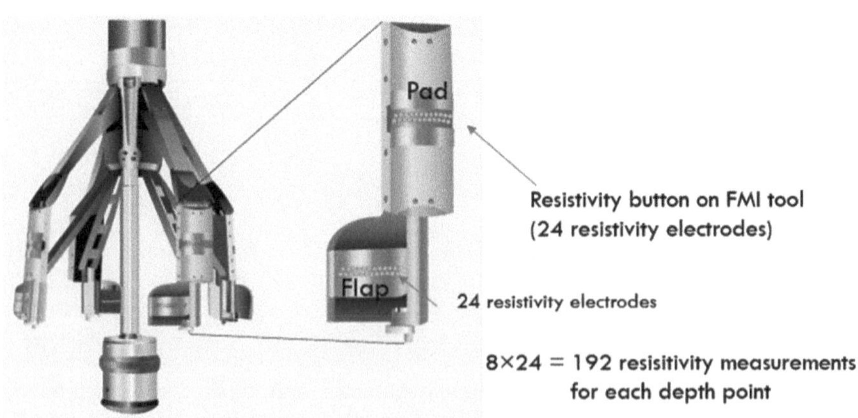

Figure 6–4 Resistivity electrodes on the Pads and Flaps of Fullbore Formation Micro-Imager (Schlumberger, 2001).

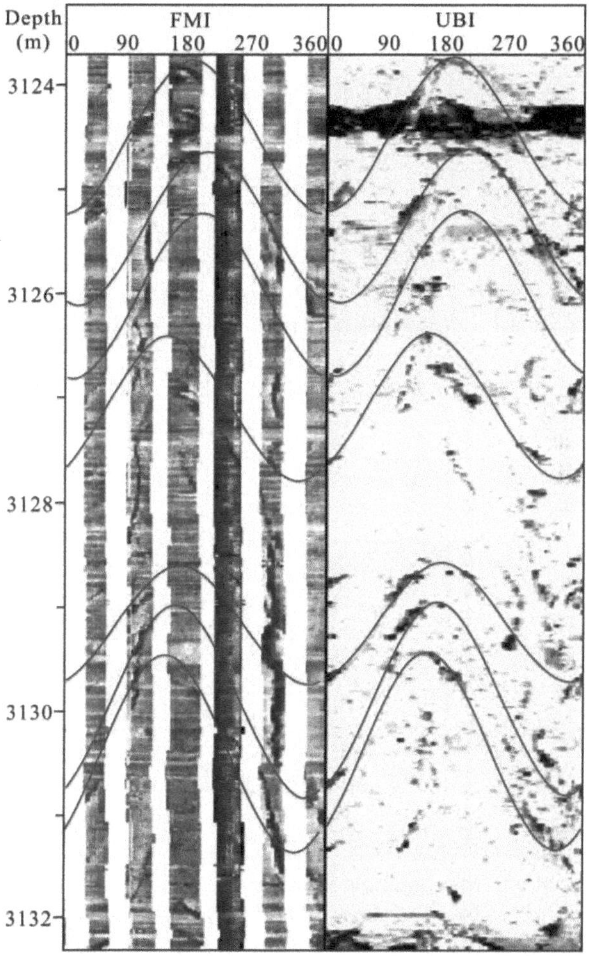

Figure 6–5 Natural fractures in Formation Micro-Imager and UBI logs from tight gas sandstones of the Xujiahe Formation, Sichuan Basin, China (Gong et al., 2019).

Fracture information such as open and filled fractures, fractures dip and azimuth, fracture spacing, and fracture intensity is derived from image logs interpretation. Induced fractures occur parallel to the direction of maximum horizontal stress, while breakouts occur parallel to the minimum horizontal stress. Analyzing the field stress map makes it possible to get clues about fractured zones and fracture development azimuths in the tight sandstone formations. On the image logs, fractures dip is generally steeper than the bed dips or structural dip. The main fracture parameters are described as follows.

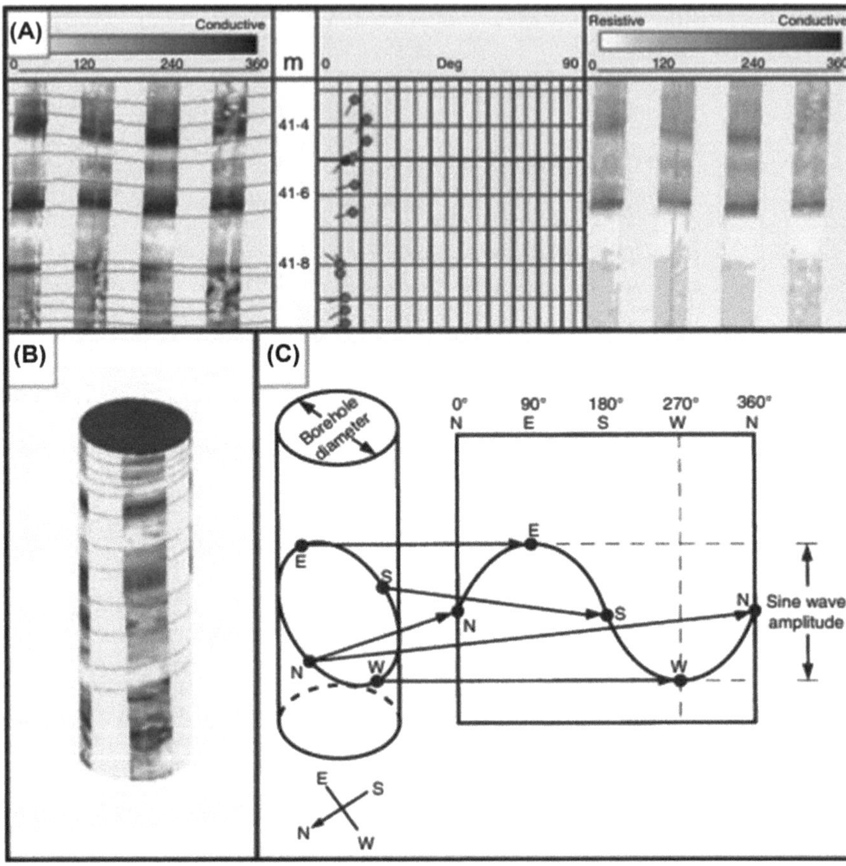

Figure 6–6 Assuming the borehole cylinder is unrolled, a fracture plane corresponds to a sinusoid. *Modified from Schlumberger, 2001. FMI Fullbore Formation MicroImager. Schlumberger Educational Services, Houston, TX.*

Fracture Dip/Strike: Each fracture is described by two clear spatial orientations, including fractures dip and strike. The angle between the horizontality and fracture plane represents fracture dip. The intersection between the fracture surface and a horizontal plane indicates fracture azimuth (Fig. 6–8). Azimuth is perpendicular to strike. Using the right-hand rule, the azimuth, strike, and dip of the fracture plane can be interpreted. If the fingers of your Right hand point down the fracture dip, then your THUMB points in the direction of the Strike. For example, 35°/45° represents a fracture plane with Strike = 35° and Dip = 45°. In terms of Azimuth/Dip, it is represented as 125°/45° (Fig. 6–8). Generally, fracture dip is steep in extensional stress regimes, while in the compressional stress

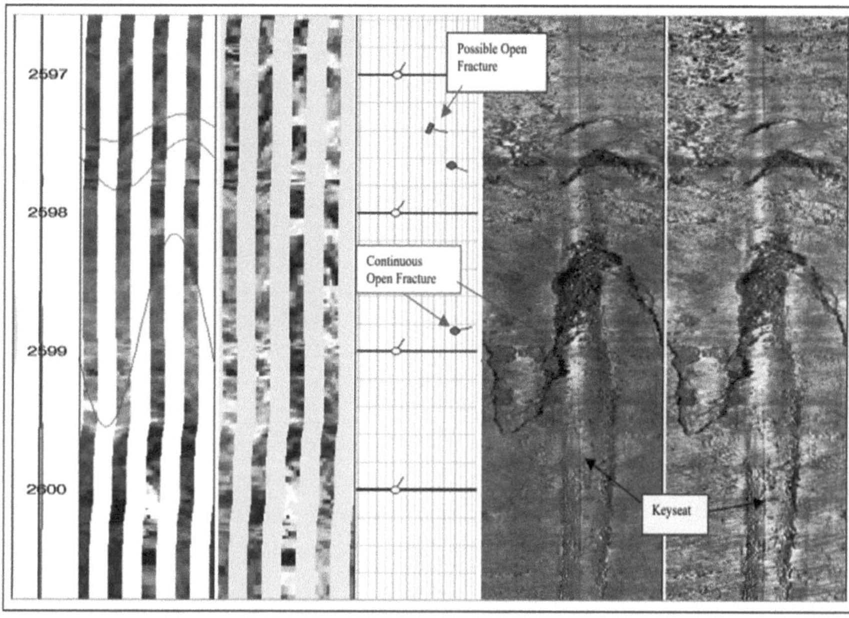

Figure 6–7 Fractures are marked by sine waves or sinusoids in the static image logs, well GS-016, GS oilfield, southwestern Iran.

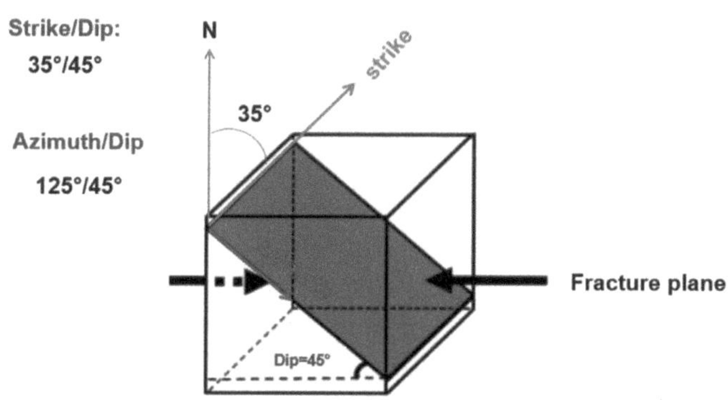

Figure 6–8 Spatial representation of a fracture plane with Strike = 35° and Dip = 45° (35°/45°). Azimuth is perpendicular to Strike (35 + 90), so in terms of Azimuth/Dip, the fracture plane is represented as 125°/45°.

system, their dip varies from high to low angles. There are different ways to show the orientation of fractures such as a Rose diagram, Stereonet, and Histogram.

Fracture Length: The total fracture length per unit length (m) or unit area (m^2) (Fig. 6–9). Having an accurate estimation of fracture length

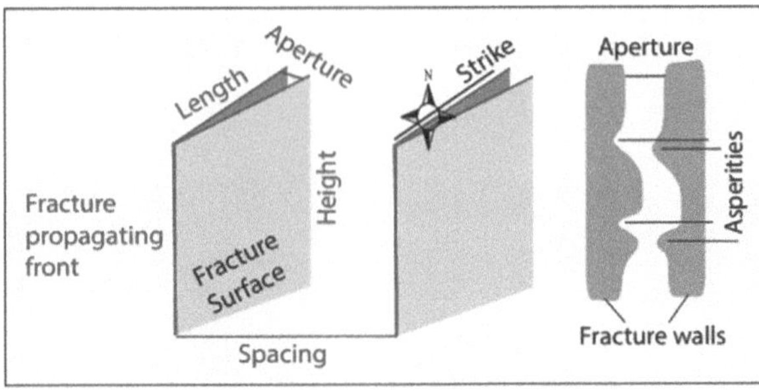

Figure 6–9 Graphical representation of fracture length, aperture, spacing, strike, height, and asperities (Sorkhabi, 2014).

is a challenging issue since it cannot be estimated from image logs or core layouts. Outcrop studies are the best reference for the estimation of maximum fracture length. The relationship between fracture aperture and fracture length can be used for this purpose. For example, fracture length is 1000 times greater than fracture aperture (Ozkaya & Mattner, 2003). Apart from this, reservoir dynamic data such as MDT results provide an idea for fracture length estimation (Hoffman et al., 2009).

Fracture Spacing: Fracture spacing (Fig. 6–9) is a measure of distance between the fractures in a fracture set along a line normal to fractures azimuth (Narr & Suppe, 1991). Brittle formations contain more fractures compared to plastic and unconsolidated formations. Thinner beds are more capable of closely-spaced fractures than thick beds. Fracture frequency is the inverse of fracture spacing (1/fracture spacing) and it simply represents the number of fractures per meter length.

Fracture Aperture: It is a measure of fracture opening and is calculated for open fractures only (Fig. 6–9). Fracture aperture is displayed in centimeters in the composite displays. The aperture of tight fractures is less than 0.25 mm, while the aperture of wide fractures is greater than 10 mm. The fracture aperture is a function of the resistivity of the drilling mud (Rm) and formation resistivity (Rxo). The oil-water emulsion used as a drilling mud well could affect the fractures' aperture in the current. However, it can still be used quantitatively to study relative changes in aperture across a tight gas formation.

$$W = c.\mathrm{A.Rm}^{\mathrm{b}}.\mathrm{Rxo}^{1-b} \tag{6.1}$$

where "W" denotes fracture aperture, "A" is the access flow of electrical current through FMI electrode; "Rm" is the resistivity of drilling-mud;

"Rxo" is the resistivity of the invaded zone; c and b are constants. Fracture aperture or spacing can be open, partially filled, or closed. Fracture filling minerals are calcite, dolomite, anhydrite, pyrite, etc.

Fracture Density: Fracture density is a measure of fracture abundance in cores, image logs, or outcrops. It can be measured in 1D, 2D, or 3D spaces. Three definitions of fracture density are as follows.

- P_{10}: the number of fractures per unit length of a rock (1/L).
- P_{20}: the number of fractures per unit area of a rock ($1/L^2$).
- P_{30}: the number of fractures per unit volume of rock ($1/L^3$).

For example, if in a 10-m interval of a tight gas sandstone, four fracture planes are identified, then the fracture density (P_{10}) is 0.4 1/m (0.12 1/ft).

Fracture Intensity: As with the fracture density, the fracture intensity is a measure of fracture abundance and can be measured in 1D, 2D, or 3D spaces. Three definitions of fracture intensity are as follows.

- P_{10}: the number of fractures per unit length of a rock (1/L).
- P_{21}: the length of fractures per unit area of a rock (1/L).
- P_{32}: the area of fractures per unit volume of a rock (1/L).

The indices 0, 1, and 2 refer to the number, length, and area of fractures, while index 3 shows the unit volume of a rock. For example, in P_{32}, 3 refers to the cube of rock, and 2 refers to the 2D fracture area. It is worth mentioning that, irrespective of P_{10}, fracture intensity is different from fracture density.

Fracture Porosity: Fracture porosity is a measure of fracture abundance and it is always defined as the ratio of fracture surface area (fracture aperture × fracture length) to the borehole area scanned by images logs. It is the type of secondary porosity caused by natural fractures openings. Fracture porosity can be measured in 1D, 2D, and 3D spaces as follows and is unitless.

- P_{11}: the thickness of fractures per unit length of a rock (L/L)
- P_{22}: the area of fractures per unit area of a rock (L^2/L^2)
- P_{33}: the volume of fractures per unit volume of a rock (L^3/L^3)

Fracture Permeability: Fracture permeability is referred to as the portion of permeability in dual-porosity reservoirs that is created by open or partially open natural fractures.

Fracture Asperity: Fractures walls are not usually smooth, parallel, and perfect. The irregularities and roughness of a fractured wall are called fracture asperity. Fracture asperity reduces fracture porosity and permeability. Having accurate knowledge of fracture asperity reduces the uncertainty in fracture flow models (Sorkhabi, 2014).

Fracture Conductivity: The multiplication of fracture width by fracture permeability ($w \times K_f$) is called fracture conductivity. It simply represents how easily fluids flow through the porous media. Dimensionless fracture conductivity (F_{cd}) is defined as the ratio of fracture conductivity over the product of formation permeability and half-length of fracture ($k \times Xf$).

6.4 Natural fractures classification

There are several classification criteria for fractures such as their geometry, origin, stress orientation, steepness, filling, aperture, spacing, and length. As is seen in Fig. 6–10, based on their geometry fractures are classified into longitudinal, transverse, diagonal (oblique), and conjugate (right-lateral or left lateral). Natural fractures can be horizontal, oblique, or vertical; however, vertical fractures are dominant and image log data can identify both. The horizontal fracture might exist for a short distance, propped open by bridging the irregular surfaces; however, most of them are sealed by overburden pressure. Unlike the induced fractures, natural fractures are usually caused by tectonic forces propagating fractures in preferential directions, characterized by a field stress regime. Fractures are in close relation to the direction of nearby folds or faults. They can be parallel, normal, and oblique to the axial plane or hinge line of the fold. They can also occur normal to the fault or in two orthogonal directions.

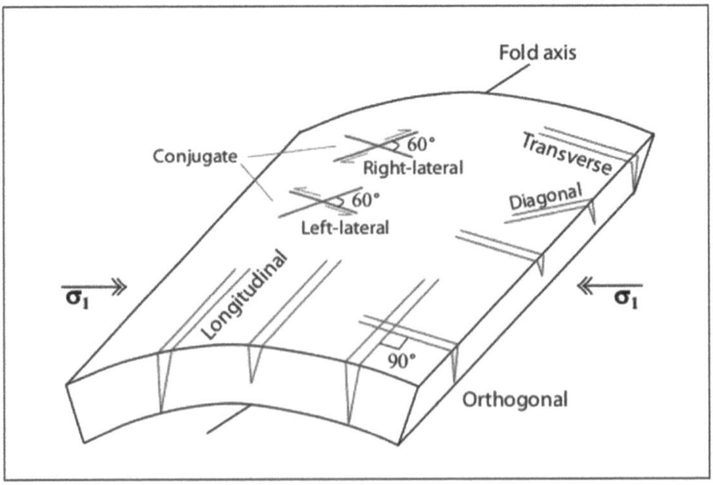

Figure 6–10 Geometrical fracture classification (Singhal and Gupta, 2010, Sorkhabi, 2014).

Based on their origin, fractures are classified into tectonic or diagenetic fractures. Tectonic fractures are relevant to the stresses orientation and field stress regime such as compressional, extensional, and strike-slip. They are caused by large-scale faulting and folding.

Diagenetic overprints such as physical and chemical compaction, mineral dissolution, and cement precipitation in the porous media alter the subsurface rock properties, pore pressures, and therefore local stresses distribution causing diagenetic fractures (Fig. 6–11).

With respect to the stresses orientation and lateral extent, fractures are further classified into four groups, including tectonic fractures, regional fractures, surface-related, and stylolites (Gudmundsson., 1987, Nelson., 2001, Laubach et al., 2004; Peacock et al., 2018). Regional fractures usually extend over very large distances. Such fractures tend to be perpendicular to the bedding surfaces and align themselves with the paleo stress direction. Stylolite fractures are solution features that occur in dolomite, limestone, and carbonate-cemented sandstone reservoirs. Stylolites are serrated planes that form perpendicular to the vertical stress as a result of pressure dissolution. Stylolites normally act as flow barriers, but if they are associated with the tensional gashes that form perpendicular to the stylolites surface they cause to enhance the reservoir quality. Surface-related fractures are visible at the surface and are mainly to due weathering along zones of weakness that may (or may not) result from the relaxation of preexisting subsurface stresses and, by definition, are not present at depth.

Figure 6–11 Photomicrograph showing a diagenetic fracture observed in a thin section from tight gas sandstones of the Xujiahe Formation, Sichuan Basin, China (Gong et al., 2019).

Geologists often base their interpretations of regional stress fields on fracture patterns seen on an outcrop. These may be quite different from what is present in the subsurface in the vicinity of hydrocarbon reservoir zones. Although fractures may be ubiquitous at the surface, there are many reasons why those fractures may not penetrate to great depths. In such situations, fractures may be nonexistent in the subsurface or simply closed by overburdened stress.

Based on their filling, fractures can be open, closed, or partially open (partially sealed). Closed fractures appear as either tight or resistive traces on image logs, while open fractures are termed as all conductive traces composed of continuous or discontinuous fractures. The conductive response of the fractures is due to the filling of fracture planes with water-based mud filtrate causing a decrease in the reading of resistivity buttons or an increase in the sonic transit time such that they appeared as dark-colored sinusoids in the FMI logs.

Based on their aperture, fractures are classified into major, medium, minor, or hairline fractures. Other open fractures such as vuggy, partially closed and partially open fractures can be considered in image logs interpretation.

Fracture spacing as a measure of distance between the parallel fracture sets can be close, moderate, and wide. There is an inverse relationship between fracture spacing and applied strain (Tang et al., 2008). The International Society for Rock Mechanics (ISRM) classifies fractures based on their spacing into seven categories (Table 6.1).

Nelson (2001) classified the naturally fractured reservoirs into four main categories with respect to porosity (matrix or fracture) and permeability (matrix or fracture) as follows (Fig. 6–12).

Table 6.1 Fracture spacing classification proposed by the International Society for Rock Mechanics.

Class	Fracture spacing (m)
Extremely close spacing	<0.02
Very close spacing	0.02–0.06
Close spacing	0.06–0.2
Moderate spacing	0.2–0.6
Wide spacing	0.6–2.0
Very wide spacing	2.0–6.0
Extremely wide spacing	>6.0

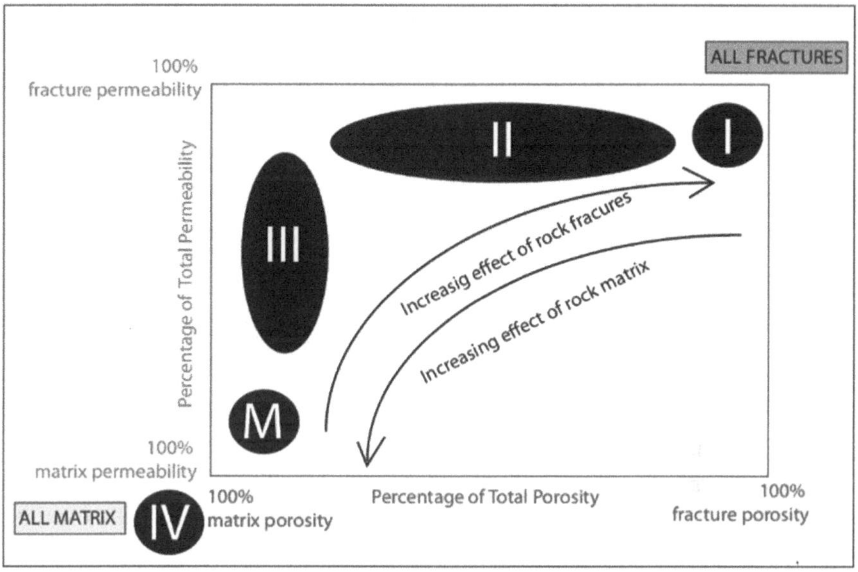

Figure 6–12 Classification of naturally fractured reservoirs. The class labels are discussed in the text. M indicates matrix permeability with no fractures and Class I shows all permeability resides in the fractures (Nelson, 2001, Sorkhabi, 2014).

- **Class I:** Fractures that are the source of porosity and permeability. This type of reservoir is common in many basins. The reservoir volume is very large and it has a very high fracture density. Production is characterized by rapid declines, and possibly early water encroachment. Its behavior approximates a single-porosity system and, thus is easy to simulate if the fracture density distribution is correctly determined and reflected in the model.
- **Class II:** Fractures facilitate most of the permeability. Although permeability is created by fractures, the porosity is provided predominantly by the reservoir's matrix. These fractured reservoirs must be simulated using dual-porosity models, especially if imbibition between matrix and fracture is the dominant flow mechanism. If not, they can be simulated by a single porosity model. Such reservoirs may show rapid fluid-production declines and early water encroachment. Enhanced recovery schemes are challenging to implement in such reservoirs because of the rapid breakthrough of the injected fluids via the fracture network, preventing them from effectively sweeping the reservoir matrix.
- **Class III:** Fractures enhance permeability, which is primarily provided by pore connectivity in the matrix. This type of reservoir is the hardest

to detect. Primary production does not usually give any indication that the reservoir is fractured. Only after fluid injection commences do unusual connectivity patterns emerge to indicate that the reservoir is affected by a natural fracture network.

- **Class IV:** Fractures do not contribute to the production and act as fluid-flow barriers.

Additionally, microfractures are worth distinguishing. They tend to be of small dimensions and either lens-shaped or display a tabular opening. Typically, microfractures require microscopic analysis to distinguish their details (Ezati et al., 2018). Microfractures can influence physical rock properties, such as strength, compressional and shear velocities, porosity, and permeability. Sometimes they can provide indications of the temperature of fracturing, fluid conditions, and timing. Many sedimentary sections provide information about the development of compaction, consolidation, and the patterns of fluid flow. Assessment of microfracture dimensions often yields key information regarding the evolution of fault zones and regional stress regimes (Gomez and Laubach, 2006; English & Laubach., 2017).

6.5 Fractures characterization from seismic data

Seismic data are the main sources of subsurface data for fracture characterization in the tight gas sandstones. Seismic attributes, as any quantitative information derived from seismic, can effectively be used for the identification of subsurface faults and fractures. Attributes can either be extracted from the prestack or poststack seismic data. Too many attributes can be extracted from seismic data but most of them are redundant information that has no relation to fractures or reservoir rock properties. Structural seismic attributes such as dip and azimuth, curvature, coherence, and ant tracking are sensitive to faults and fracture.

It was shown that horizon-based dip and azimuth attributes are very useful in delineating subtle faults and fractures (Rijks and Jauffred, 1991) if their displacement is a function of a seismic wavelet. Curvature and coherence attributes have widely been used for fractures identification. The curvature attribute measures the lateral changes in azimuth and dip. The coherence attribute was introduced by Bahorich and Farmer (1995) to compute discontinuity in seismic data using a cross-correlation algorithm. It somehow measures the lateral changes in the waveform to characterize the faults and fractures. There are many curvature attributes from which the most positive, the most negative, and Gaussian curvatures are more correlated with faults and fractures. More accurate coherence calculation was achieved through semblance and eigenstructure to highlight the structural dip (Marfurt et al., 1999).

Ant tracking is an illumination attribute that can be used for faults and fractures identification. In this approach, a lot of artificial ants are created on seismic data such that the ants will move along the fracture or fault planes and will leave pheromone.

6.6 Curvature analysis

The curvature analysis is an instrumental way of faults and fractures characterization in the structures hosting tight gas sandstones. The curvature is the reciprocal of the radius of a hypothetical circle fitted to a geological structure in a given direction. It Simply means how sharply a bent structure deviates from being a straight surface. Mathematically, the curvature is the second derivative of topography that can be calculated in all directions. The inputs to perform the curvature analysis are structural horizons derived from seismic data or interpolation of well tops. Curvature analysis delivers a predictive map of the large-scale structural features that could be used if such fractures are proven to impact the reservoir, to generate an equivalent fracture permeability model (Stewart & Podolski, 1998). In a curvature map, high curvature areas correspond to the location of faults and fractures.

The maximum curvature (k_{max}): Also called the first principal curvature, is the measure of the maximum curvature at a given point. In practice, the second derivative of the surface is computed for any point on the surface in all directions from 0° to 360°. At each point, k_{max} is set to the maximum computed value. The direction corresponding to k_{max} is also kept in memory as the direction of the first principal curvature (this direction can sometimes be used to predict fracture orientation).

The minimum curvature (k_{min}): The methodology to calculate the minimum curvature is similar to that of the maximum curvature, but the minimum values in all directions are kept. The azimuth of maximum curvature is orthogonal to the minimum curvature azimuth.

The Gaussian curvature: Gaussian curvature is defined as the product of minimum curvature and maximum curvature. Directional curvature maps are computed in a specific direction to highlight the specific trends. A graphical illustration representing the maximum, minimum, and Gaussian curvature is shown in Fig. 6–13. For a folded surface with a North-South

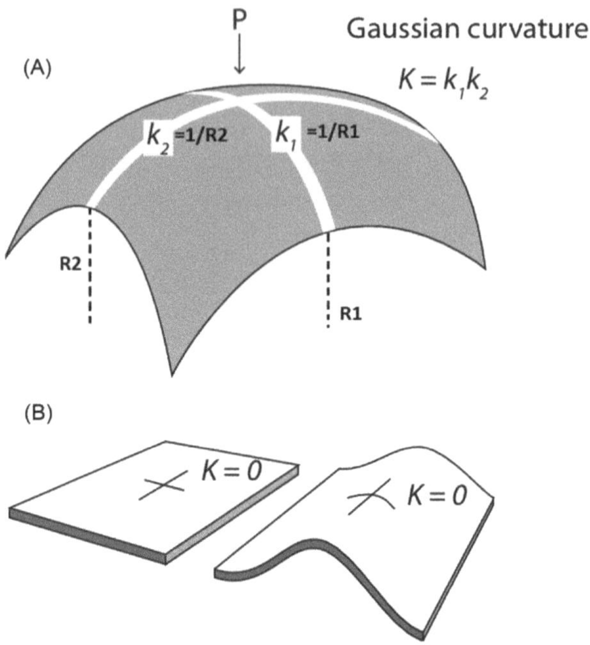

Figure 6–13 Graphical illustration representing maximum (k_1), minimum (k_2), and Gaussian curvature (K). (a) Gaussian curvature is defined as the product of maximum and minimum curvature ($1/R_1 \times 1/R_2$). (b) Gaussian curvature for both lower shapes, either curved plate ($k_1 > 0$, $k_2 = 0$) and smooth plate ($k_1 = 0$, $k_2 = 0$) is equal to zero (Cosgrove, 2015; Paragoda and Toda 2016) (k_1: maximum curvature, k_2: minimum curvature, R_1: radius in maximum curvature direction, R_2: radius in minimum curvature direction). *After Paragoda, T., Toda, M., 2016 Willmore and generalized willmore energies in space forms (Ph.D. dissertation). Texas Tech University, pp. 90.*

axis, the maximum curvature corresponds to the East-West direction, while the curvature is zero in the North-South direction (minimum curvature). The product of the two curvatures is termed the Gaussian curvature (Wynn and Stewart, 2003). As is seen in Fig. 6–13, on the crest of the structure, the Gaussian curvature is therefore zero. Accordingly, this type of curvature cannot detect extrados fractures that are denser on the fold crests. Gaussian curvature is a qualitative way to assess if without deformation a structure can be unfolded. It can be used to interpret highly fractured zones in a tightly folded structure. Curvature intensity is considered positive when concavity is turned toward the top and negative when concavity is turned toward the bottom (Fig. 6–14). Subsurface faults detection through the curvature map with an example from the top Sue Formation in the Whicher Range tights gas field of the Perth Basin (Australia) is shown in

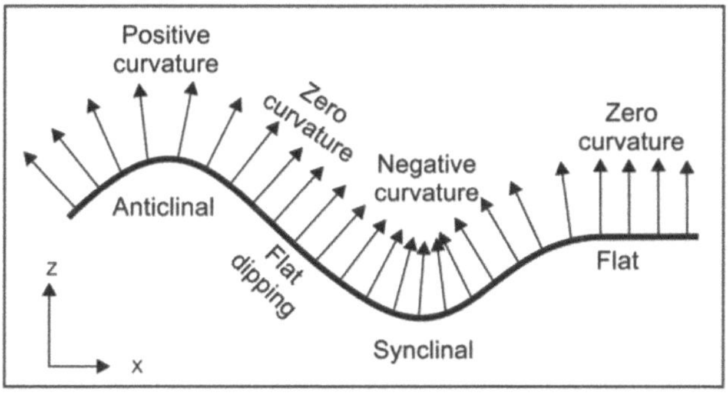

Figure 6–14 Definition of curvature sign. Anticlines have positive curvature, synclines have negative curvatures and flat surfaces show zero curvature (Roberts, 2001; Francelino and Antunes, 2013). *Reproduced with permission from Roberts, A., 2001. Curvature attributes and their application to 3D interpreted horizons. First Break, 18: 85–99; Francelino, F., Antunes, A., 2013. Applying filters and seismic attributes for enhancing faults in the 3D seismic survey of Alto de Siririzinho (Sergipe-Alagoas Basin, northeast Brazil). Revista Brasileira de Geofísica 31(1), 109–123.*

Fig. 6–15. As is seen, the curvature map shows a good agreement with the superimposed faults of the structure.

6.7 The role of natural fractures on tight gas field production

Production from the low permeability tight gas sandstones relies on natural fractures development. For the practical analysis and development of such reservoirs, we need to recognize the role of naturally forming fractures in the different reservoir development stages. To study the impact of natural fractures on production, a multistep modeling process needs to be adopted. Accordingly, fracture characterization through the integration of multiple sources of geosciences and reservoir engineering data plays a pivotal role in a deep understanding of the reservoir production mechanisms and provides a more realistic dynamic simulation model of tight gas sandstones. Data to build fracture models come from image logs, conventional good logs, cores, and seismic data. To make a fracture network model, first, a Continuous Fracture Network (CFN) needs to be created based on the fracture intensity data. Normally, data to make fracture intensity model come from image logs interpretation. All interpreted fracture planes are used for fracture density modeling. Afterward, a Discrete Fracture Network (DFN) model is made to represent fracture properties for

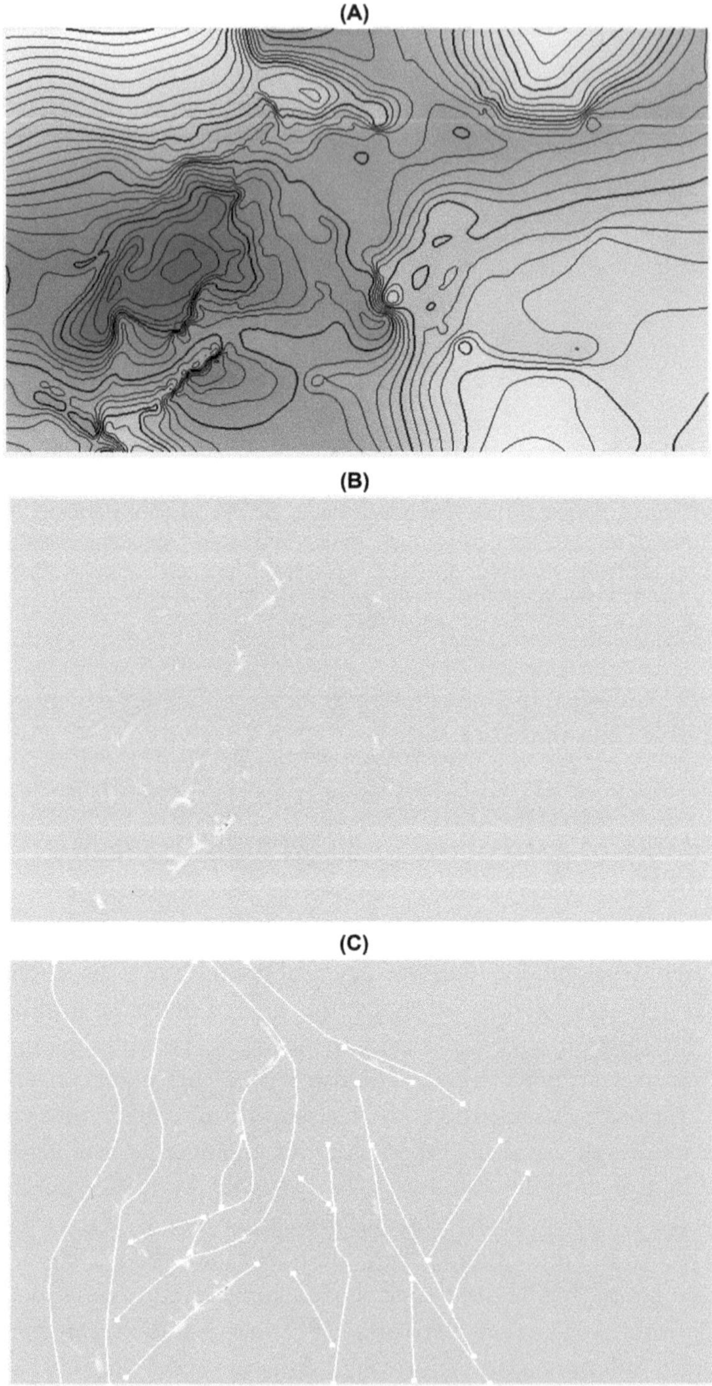

Figure 6–15 Plots showing identification of subsurface faults through structural curvature map for the top Sue Formation, Whicher Range gas field, Perth Basin. Underground contour map of top Sue Formation (A) and its extracted maximum curvature map (B). The curvature map shows a good agreement with the superimposed structural faults (C).

hydraulically connected fractures. The DFN is a group of planes in 3D space representing fractures. Each fracture network is made up of at least one fracture set but it may contain more sets.

It should be pointed out that the 3D model of a DFN is constructed stochastically through numerical propagation of fracture attributes in a 3D grid. Meanwhile, such properties can vary in 3D space and can easily be modeled using seismic attributes. The final results of stochastic DFN modeling will be upscaled to a simulation grid in a dynamic model. Four fracture parameters affecting the fluid flow and dynamic simulation results are fracture porosity, fracture permeability, fracture length, and sigma parameter. Accordingly, the effect of the fracture model can be investigated in the dynamic simulation model. The production data can be correlated to the fracture intensity. Productivity index (PI), expressed in STB/D/psi, is an important well parameter showing the oil or gas well's ability to produce. It is defined as the ratio of the total flow rate of liquid at the surface to the pressure drawdown at the middle of the producing depth interval. PI corresponds to a formation thickness that produces. It is important therefore to determine this thickness to characterize the deliverability of a given well appropriately. Typically, the higher intensity of fractures corresponds to the high-producing wells. In this regard, the integration of fracture data from core and image logs aids in locating fractured zones with high production rates mostly located adjacent to the anticline crest. Mud loss data provide very useful proxies for fracture characterization of tight gas reservoirs. Intervals with the high mud loss rates are well correlated to the fractured zones. A strong agreement is always seen between the faults spatial model, curvature map, mud loss, fracture density, and productivity index.

6.8 Mechanical and fracture stratigraphy

Within a relatively undeformed geological formation, fracture development patterns such as fracture spacing, length, and intensity are highly controlled by mechanical stratigraphy rather than folds and faults (Underwood et al., 2003). Mechanical stratigraphy is related to geological formations with nearly similar or consistent mechanical properties. The mechanical interface and mechanical units are two main topics considered in mechanical stratigraphy. A mechanical interface is referred to as a boundary or bed with a certain thickness within which fractures propagation is terminated. A mechanical unit can be composed of a single lithology unit or multiple lithologies with similar geomechanical

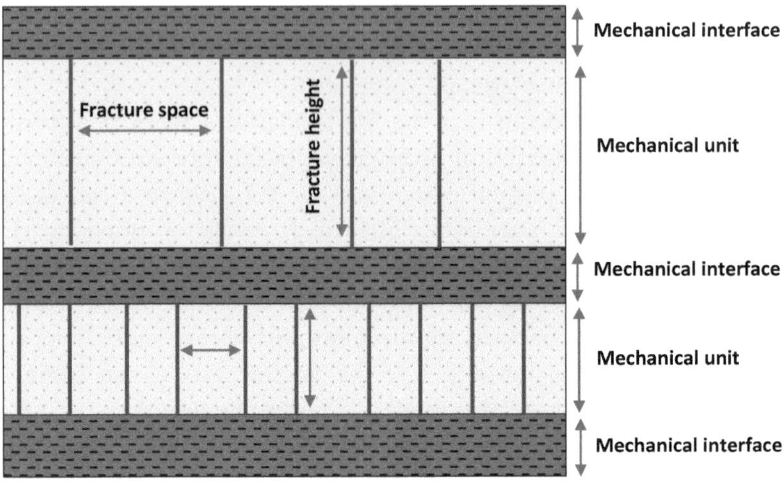

Figure 6–16 A schematic model of mechanical stratigraphy concept. A set of vertical fractures are developed in the upper and lower tight sandstone formations and shaly interbeds act as mechanical interfaces terminating fractures propagation.

properties. It can be said that fractures propagate within the mechanical units and are terminated by the mechanical interfaces (Cooke et al., 2006; Laubach et al., 2009; Rustichelli et al., 2013). Fracture stratigraphy subdivides the stratified formations into discrete fracture units based on fracture attributes such as fracture length, intensity, and spacing, while mechanical stratigraphy classifies the stratified formations into discrete mechanical units with similar elastic properties, brittleness index, and tensile strength (Laubach et al., 2009). Some authors studied the impact of lithofacies and stratigraphic architecture on fracture and mechanical stratigraphy in sandstone and carbonate formations (e.g., Barbier et al., 2012; Giuffrida et al., 2020). On the other hand, the controlling role of mechanical stratigraphy on fracture patterns development in tight gas sandstones is understudied and needs more investigation. A schematic model of mechanical stratigraphy is illustrated in Fig. 6–16. As is seen, vertical fractures are developed in two tight sandstone formations (mechanical units) and shaly interbeds act as mechanical interfaces terminating fractures propagation.

References

Awdal, A., Healy, D., Alsop, G.I., 2016. Fracture patterns and petrophysical properties of carbonates undergoing regional folding: a case study from Kurdistan. N Iraq Journal of Marine and Petroleum Geology 71, 149–167.

Bahorich, M.S., Farmer, S.L., 1995. 3-D seismic discontinuity for faults and stratigraphic features: the coherence cube. The Leading Edge 14, 1053–1058.

Barbier, M., Hamon, Y., Callot, J.P., Floquet, M., Daniel, J.M., 2012. Sedimentary and diagenetic controls on the multiscale fracturing pattern of a carbonate reservoir: the madison formation (Sheep Mountain, Wyoming, USA). Marine and Petroleum Geology 29, 50–67.

Cooke, M.L., Simo, J.A., Underwood, C.A., Rijken, P., 2006. Mechanical stratigraphic controls on fracture patterns within carbonates and implications for groundwater flow. Sedimentary Geology 184 (3), 225–239.

Corbett, K., Friedman, M., Spang, J., 1987. Fracture development and mechanical stratigraphy of Austin Chalk, Texas. American Association of Petroleum Geologists Bulletin 71, 17–28.

Cosgrove, J., 2015. The association of folds and fractures and the link between folding, fracturing, and fluid flow during the evolution of a fold-thrust belt: a brief review. Geological Society 421 (1), London Special Publications.

English, J.M., Laubach, S.E., 2017. Opening-mode fracture systems- insights from recent fluid inclusion microthermometry studies of crack-seal fracture cements. In: Turner, J. P., Healy, D., Hillis, R.R., Welch, M. (Eds.), Geomechanics and Geology, Geological Society. Special Publications, London, p. 458.

Ezati, M., Azizzadeh, M., Riahi, M.A., Fattahpour, V., Honarmand, J., 2018. Characterization of micro-fractures in carbonate Sarvak reservoir, using petrophysical and geological data, SW Iran. Journal of Petroleum Science and Engineering 170, 675–695.

Francelino, F., Antunes, A., 2013. Applying filters and seismic attributes for enhancing faults in the 3D seismic survey of Alto de Siririzinho (Sergipe-Alagoas Basin, northeast Brazil). Revista Brasileira de Geofisica 31 (1), 109–123.

Giuffrida, A., Agosta, F., Rustichelli, A., Panza, E., La Bruna, V., Eriksson, M., et al., 2020. Fracture stratigraphy and DFN modelling of tight carbonates, the case study of the lower cretaceous carbonates exposed at the Monte Alpi (Basilicata, Italy). Marine and Petroleum Geology 112, 104045.

Gomez, L.A., Laubach, S.E., 2006. Rapid digital quantification of microfracture populations. Journal of Structural Geology 28 (3), 408–420.

Gong, L., Su, X., Gao, S., Fu, X., Jabbari, H., Wang, X., et al., 2019. Characteristics and formation mechanism of natural fractures in the tight gas sandstones of Jiulongshan gas field, China. Journal of Petroleum Science and Engineering 175, 1112–1121.

Gudmundsson, A., 1987. Geometry, formation and development of tectonic fractures on the Reykjanes Peninsula, southwest Iceland. Tectonophysics 139 (3–4), 295–308.

Gutmanis, J., Ardèvol i Oró, L., Díez-Canseco, D., Chebbihi, L., Awdal, A., Cook, A., 2018. Fracture analysis of outcrop analogues to support modelling of the subseismic domain in carbonate reservoirs, south-central Pyrenees. In: Ashton, M., Dee, S.J., Wennberg, O.P. (Eds.), Subseismic-Scale Reservoir Deformation, Geol. Soc., 459. Spec. Pub., London, pp. 139–156.

Hoffman, B.T., Narr, W., Li, L., 2009, Using PLT data to estimate the size of natural fractures. In: SPE Paper 121293, the SPE Western Regional Meeting, 24–26 March.

Laubach, S.E., Olson, J.E., Gross, M.R., 2009. Mechanical and fracture stratigraphy. AAPG Bulletin 93 (11), 1413–1426.

Laubach, S.E., Reed, R.M., Olson, J.E., Lander, R.H., Bonnell, L.M., 2004. Coevolution of crack-seal texture and fracture porosity in sedimentary rocks: cathodoluminescence observations of regional fractures. Journal of Structural Geology 26 (5), 967–982.

Marfurt, K.J.V., Sudhaker, A., Gersztenkorn, K.D., Crawford, Nissen, S.E., 1999. Coherence calculations in the presence of strong structural dip. Geophysics 64, 104–111.

Narr, W., Suppe, J., 1991. Joint spacing in sedimentary rocks. Journal of Structural Geology 13, 1037–1048.

Nelson, R., 2001. Geologic Analysis of Naturally Fractured Reservoirs. Elsevier, p. 352.
Ozkaya, S.I., Mattner, J., 2003. Fracture connectivity from fracture intersections in borehole image logs. Computers & Geosciences 29 (2), 143–153.
Paragoda, T., Toda, M., 2016 Willmore and generalized willmore energies in space forms (Ph.D. dissertation). Texas Tech University, pp. 90.
Peacock, D.C.P., Sanderson, D.J., Rotevatn, A., 2018. Relationships between fractures. Journal of Structural Geology 106, 41–53.
Pollard, D.D., Aydin, A., 1988. Progress in understanding jointing over the past century. Geological Society of America Bulletin 100, 1181–1204.
Price, N., 1966. Fault and Joint Development in Brittle and Semi-Brittle Rock. Pergamon Press, Oxford, p. 176.
Price, N., Cosgrove, J., 1990. Analysis of Geological Structures. Cambridge University Press, Cambridge, p. 502.
Rijks, E.J.H., Jauffred, J.C.E.M., 1991. Attribute extraction: an important application in any detailed 3-D interpretation study. The Leading Edge 10, 11–19.
Roberts, A., 2001. Curvature attributes and their application to 3D interpreted horizons. First Break 18, 85–99.
Rustichelli, A., Agosta, F., Tondi, E., Spina, V., 2013. Spacing and distribution of bed perpendicular joints throughout layered, shallow-marine carbonates (Granada Basin, southern Spain). Tectonophysics 582, 188–204.
Schlumberger, 2001. FMI Fullbore Formation MicroImager. Schlumberger Educational Services, Houston, TX.
Singhal, B.B.S., Gupta, R.P., 2010. Applied Hydrogeology of Fractured Rocks. Springer, p. 344.
Sorkhabi, R., 2014. Fracture, fracture everywhere. GeoExPro Magazine 11 (3), 86–88.
Stewart, S.A., Podolski, R., 1998. Curvature analysis of gridded geological surfaces. In: Coward, M.P., Daltaban, T.S., Johnson, H. (Eds.), Structural Geology in Reservoir Characterization, Geological Society, 209. Special Publication, London, pp. 133–147.
Tang, C.A., Liang, Z.Z., Zhang, Y.B., Chang, X., Tao, X., Wang, D.G., et al., 2008. Fracture spacing in layered materials: a new explanation based on two-dimensional failure process modeling. American Journal of Science 308 (1), 49–72. January.
Underwood, C.A., Cooke, M.L., Simo, J.A., Muldoon, M.A., 2003. Stratigraphic controls on vertical fracture patterns in Silurian dolomite, northeastern Wisconsin. AAPG Bulletin 87 (1), 121–142.
Wynn, T., Stewart, S.A., 2003. The role of spectral curvature mapping in characterizing subsurface strain distributions. Eological Society 209 (1), 127–143. London Special Publications.
Zhou, X., Xu, G., Cui, H., Zhang, W., 2020. Fracture development and hydrocarbon accumulation in tight sandstone reservoirs of the Paleogene Huagang Formation in the central reversal tectonic belt of the Xihu Sag, East China Sea. In: Petroleum Exploration and Development 47 (3), 499-512.

CHAPTER 7

Tight gas petrophysical evaluation

7.1 Introduction

Petrophysical evaluation of well logging data is a pivotal step in reservoir evaluation of tight gas reservoirs. Petrophysical parameters such as porosity, the volume of clay, water saturation, permeability, pay zones and gas-water contact are obtained from the petrophysical evaluation of well logs. Considering the tight nature of these reservoirs, and accurate petrophysical analysis is very critical for their development and economic evaluation. For example, in a conventional reservoir with a porosity of 20%, an error of 2 p.u. in porosity estimation is not a great concern. The difference between 20% and 18% will not greatly affect volumetric evaluation and decision makings. However, the same error of 2 p.u. in a tight gas reservoir of 5% average porosity can lead to significant errors in prospect evaluation and development. Both deterministic and probabilistic models can be used for the interpretation of well logs in tight sands. Deterministic approaches are done in a step-by-step manner so that the output of the former step is the input for the next step. Probabilistic models are employed in case of having a complete set of data from the studied reservoir including a full set log, core results (petrographic data, CCAL, SCAL, etc.), and unconventional logs (e.g., nuclear magnetic resonance (NMR) and elemental capture spectroscopy) such that a large number of petrophysical properties can simultaneously be estimated through solving multi-mineral equations.

7.2 Deterministic method

7.2.1 Volume of shale

The volume of shale is calculated from natural gamma-ray tools (NGT) or natural gamma-ray spectroscopy (NGS). NGT logs include CGR and SGR logs. CGR measures the radioactive thorium and potassium concentrations, while SGR measures uranium in addition to ^{232}Th and ^{40}K. NGS logs

Reservoir Characterization of Tight Gas Sandstones
DOI: https://doi.org/10.1016/B978-0-323-90180-2.00011-6

include ^{232}Th, ^{40}K, and ^{238}U logs. Clay minerals and shale due to their small pore types system can trap high amounts of the radioactive elements. The gamma-ray log is represented on a scale of 0 (right limit) to 200 (left limit). Deflection of gamma-ray to the left side (zero) means clean formation, while its deflection to the right side (200 API) indicates shaly formation. A graphical example of the volume of shale calculation using a gamma-ray log is shown in Fig. 7–1. It should be noted that Arkosic sandstones are high in GR due to the presence of potassium in the structure of their minerals and should not be confused with shales or clays.

In order to calculate the volume of shale from GR logs, the first gamma-ray index needs to be calculated by using minimum, maximum, and depth reading through one of the following equations.

$$I_{GR} = \frac{GR - GR_{min}}{GR_{max} - GR_{min}} \tag{7.1}$$

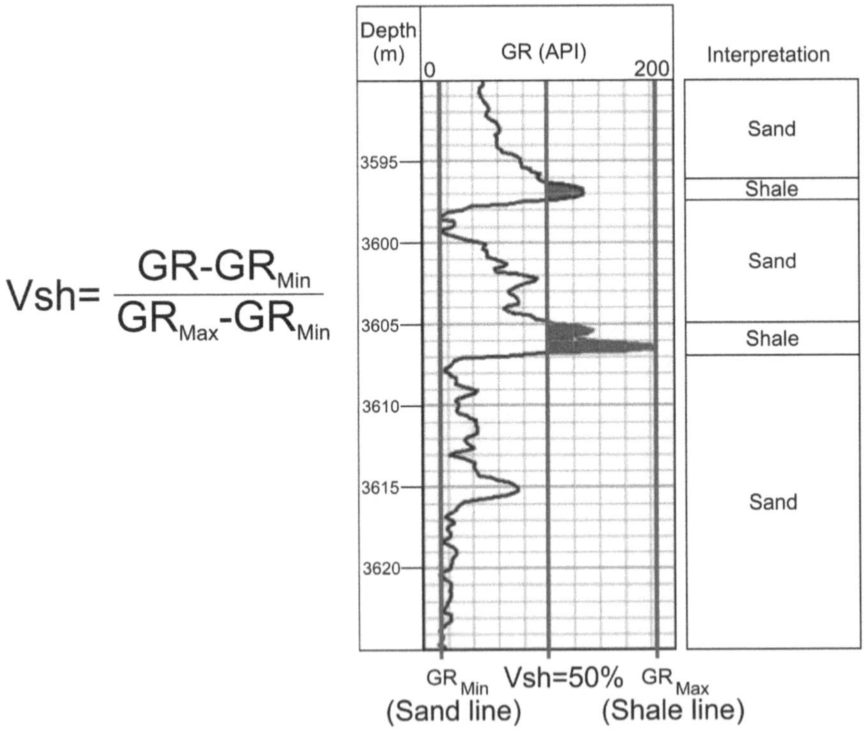

Figure 7–1 A sample of gamma-ray was used to discriminate between shale and sandstones. *Originally generated for this chapter.*

$$I_{GR} = \frac{CGR - CGR_{min}}{CGR_{max} - CGR_{min}} \tag{7.2}$$

$$I_{GR} = \frac{SGR - SGR_{min}}{SGR_{max} - SGR_{min}} \tag{7.3}$$

$$I_{GR} = \frac{^{40}K - {}^{40}K_{min}}{^{40}K_{max} - {}^{40}K_{min}} \tag{7.4}$$

$$I_{GR} = \frac{^{232}Th - {}^{232}Th_{min}}{^{232}Th_{max} - {}^{232}Th_{min}} \tag{7.5}$$

Calculating the gamma-ray index, the volume of shale can be derived using one of the following Eqs. (7.6)–(7.9).

Larionov's (1969) equation for tertiary rocks:

$$V_{sh} = 0.083 \times \left(2^{3.7058 I_{GR}} - 1\right) \tag{7.6}$$

Larionov's (1969) equation for older rocks:

$$V_{sh} = 0.33 \times \left(2^{2 I_{GR}} - 1\right) \tag{7.7}$$

Clavier et al. (1971) equation

$$V_{sh} = 1.7 - \left(3.38 - (I_{GR} + 0.7)^2\right)^{0.5} \tag{7.8}$$

Stieber (1970) equation

$$V_{sh} = \frac{3 I_{GR}}{(1 + 2 I_{GR})} \tag{7.9}$$

The Spontaneous potential (SP) log can also be used to calculate the volume of shale. SP measures the SP between a fixed electrode located on the ground surface and another electrode that is moved into a well. This potential, which varies from formation to formation, ranges from a few tens to a few hundred millivolts. The main causer of the SP log is the difference between mud filtrate resistivity (R_{mf}) and formation water resistivity (R_w) (liquid junction or diffusion potential). The greater the difference between R_{mf} and R_w, the greater the response of SP log (the greater the separation between the shale and sand lines) and vice versa. If the formation water resistivity is equal to the drilling mud filtrate resistivity $R_{mf} = R_w$), the SP log does not have a response (zero millivolt potential, no deflection to the right or left). If $R_{mf} > R_w$, then the deflection of SP to the left indicates sands, and its deflection to the right shows shales (Fig. 7–2A). If $R_{mf} < R_w$, then the deflection of SP to the left indicates shales, and its deflection to the right shows sands (Fig. 7–2B). Using the SP log, the volume of shale is calculated as follows.

$$V_{sh} = \frac{SP - SP_{min}}{SP_{max} - SP_{min}} = \frac{PSP - SSP}{SP_{shale} - SSP} \tag{7.10}$$

$$V_{sh} = 1 - \frac{PSP}{SSP} \tag{7.11}$$

where SP, SP_{min}, SP_{max} and are the SP log reading against the depth of interest, maximum and minimum values of SP log. Potential SP (PSP) is the deflection between sand line and shale line for an interval of interest and Static SP (SSP) is the maximum deflection between sand line and shale line.

For example, to calculate the volume of shale based on SP log ($R_{mf} > R_w$) shown in Fig. 7–3, the first sand line and shale line are drawn. The shale line corresponds to the maximum deflection of the SP log to the right and

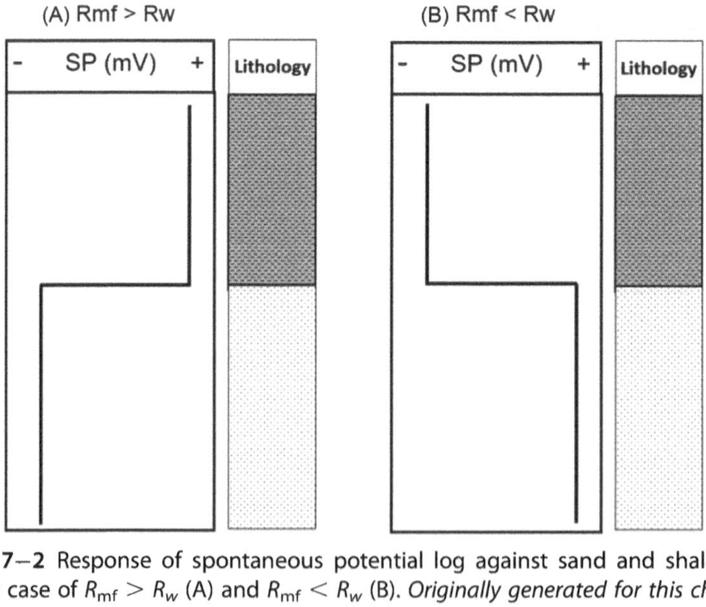

Figure 7–2 Response of spontaneous potential log against sand and shale formations in case of $R_{mf} > R_w$ (A) and $R_{mf} < R_w$ (B). *Originally generated for this chapter.*

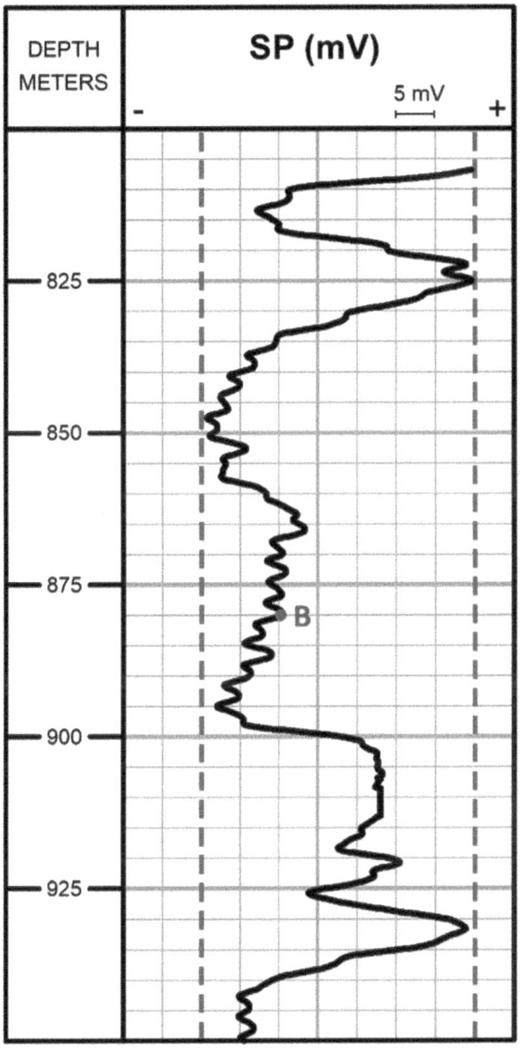

Figure 7–3 A sample of spontaneous potential log. *Originally generated for this chapter.*

represents the zero potential. According to Fig. 7–3, in the SP log header, each unit is given 5 mV as the horizontal axis scale, so the maximum deviation from the shale line to the left is −35 mV. At point A, the deviation of the SP log from the shale line is −25 mV (Fig. 7–3). So we have,

$$V_{sh} = \frac{SP_{log} - SP_{clean}}{SP_{shale} - SP_{clean}} = \frac{-25 - (-35)}{0 - (-35)} = 0.29$$

or using Eq. (7.11),

$$V_{sh} = 1 - \frac{\text{PSP}}{\text{SSP}} = 1 - \frac{-25}{-35} = 0.29$$

Therefore, in both methods, the amount of shale volume at a depth of 880 m is equal to 29%.

As shown in Fig. 7–4, different types of clays occur in sandstone reservoirs including dispersed clay (pore filling, pore-lining, and pore bridging), laminated clay, and structural clay (rock fragments, rip-up clast, and clay-replaced grains).

7.2.2 Porosity

Although the main application of porosity logs is to calculate reservoir porosity, their interpretation individually or in the form of composite logs provides an effective way for tight gas intervals detection and evaluation. Three main porosity logs include neutron, sonic, and density. In this section, the first porosity calculation using the conventional methods is explained in brief. Afterward, the methods for tuning and calibration of well log-derived porosity to the core porosity are discussed to have an accurate porosity estimation in tight sandstones.

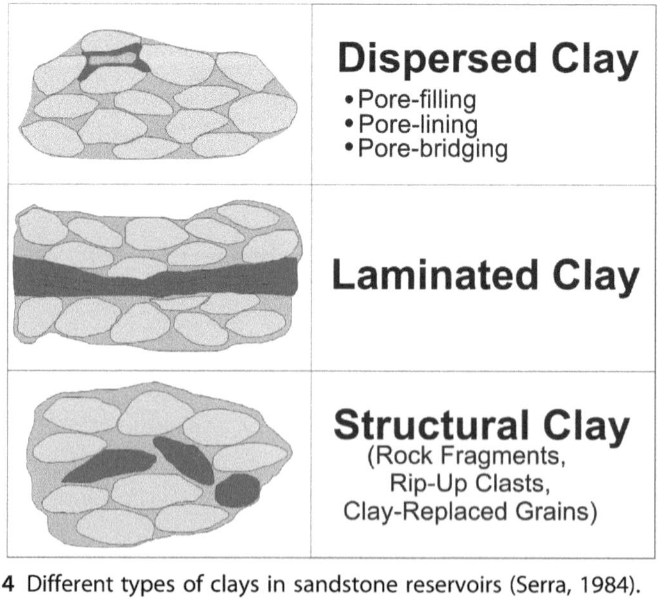

Figure 7–4 Different types of clays in sandstone reservoirs (Serra, 1984).

The neuron log (NPHI) measures the formation porosity based on the concentration of hydrogen atoms in the formation. When the high-energy neutrons collide with the atoms of the equal mass, their energy drops, quickly. Neutron particles have the same mass as hydrogen atoms. Thus, the higher the concentration of the hydrogen atoms in rocks, the higher the energy drop; resulting in a higher formation porosity reading and vice versa. As the neutron log measures the formation porosity as a function of hydrogen atoms concentration, its response will be weaker in gas-bearing formation. The hydrogen concentration in gas is less than in oil and water causing an underestimation of formation porosity in gas zones. After processing and conversion of neutron counts to porosity, the neutron log is represented between −0.15 p.u. (right) and 0.45 p.u. (left). A sample of neutron log is shown in Fig. 7−5. Neutron log should be corrected for borehole environmental effects, gas effect, matrix effect, and shale effect. As the neutron tool is calibrated for limestone rock, it needs to be corrected for sandstones. Normally, in sandstones 4 p.u. needs to be added to the NPHI log to account for the matrix effect. The presence of shale as a laminate, scattered, or structural leads to an increase in neutron log readings and overestimation of the actual porosity. The shale-corrected neutron porosity

(φ_{Nshc}) is calculated using Eq. (7.12).

$$\varphi_{\mathrm{Nshc}} = \varphi_{\mathrm{Na}} - (V_{\mathrm{sh}} \times \varphi_{\mathrm{Nsh}}) \tag{7.12}$$

where φ_{Na} is apparent neutron porosity derived from the NPHI log, V_{sh} is a volume of shale, and φ_{Nsh} is the reading of neutron log against a shale formation.

Sonic log (DT) measures the time required for transmission of acoustic waves from a foot of the formation. The sonic log is directly correlated to formation porosity, the higher the sonic transit time, the higher the formation porosity. Sonic log reading increases significantly in gas zones causing sharp peaks called cycle skipping. Cycle skipping can also be an indicator of fractures in the formation and hence they must not be mistaken for gas effect. Interpretation of composite logs such as neutron density can distinguish fractures from gas-bearing zones. The sonic log is represented between 40 (right) and 140 μs/f (left) (Fig. 7−6). The deviation of the sonic log to the right indicates low porosity formation and its deviation to the left indicates high porosity formation. Porosity is calculated from the sonic log (φ_S) using Wyllie's equation as follows.

$$\varphi_S = \frac{\mathrm{DT}_{\log} - \mathrm{DT}_{\mathrm{ma}}}{\mathrm{DT}_f - \mathrm{DT}_{\mathrm{ma}}} \tag{7.13}$$

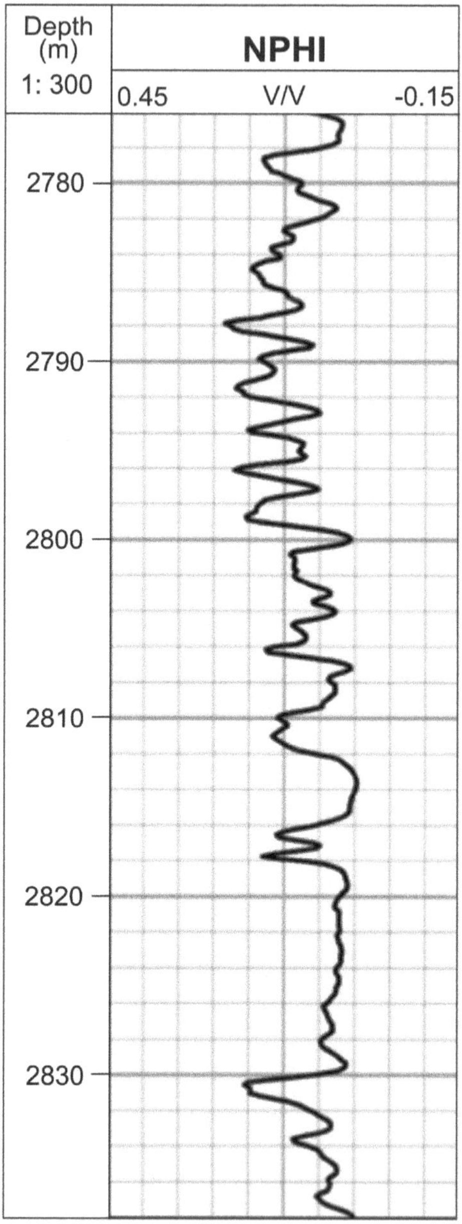

Figure 7–5 A sample of neutron log. The deviation of the neutron log to the right indicates a lower porosity and its deviation to the left indicates a higher porosity of the formation. Neutron porosity in anhydrides tends to be −0.15. *Originally generated for this chapter.*

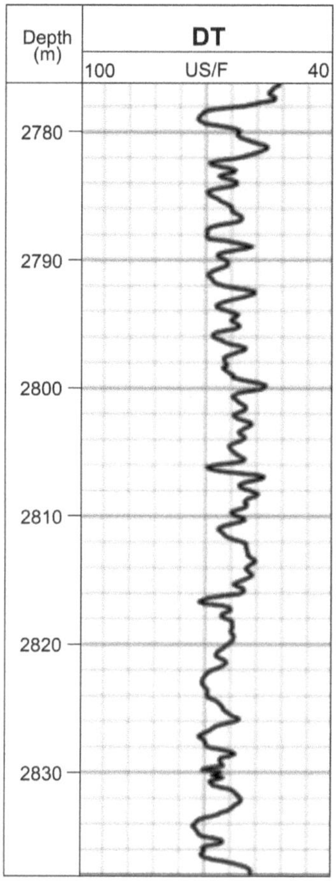

Figure 7–6 A sample of the sonic log. The deviation of the sonic log to the right indicates compact formation and its deviation to the left indicates porous formation. *Originally generated for this chapter.*

where DT_{log} is the sonic log reading at each depth point, DT_{ma} is the sonic transit time from matrix and DT_f is the sonic transit time from fluid (189.5 μs/f for the case of brine). DT_{ma} is taken as 55, 65, 49, and 44 μs/f for shale, sandstone, limestone, and dolomite, respectively.

Roughly, to correct the gas effect on sonic porosity, Eq. (7.14) is used.

$$\varphi_{Scorr} = 0.7 \times \varphi_S \tag{7.14}$$

The shale-corrected sonic porosity (φ_{Sshc}) is calculated using Eq. (7.15).

$$\varphi_{Sshc} = \varphi_S - (V_{sh} \times \varphi_{Ssh}) \tag{7.15}$$

where φ_S is apparent sonic porosity derived from Wyllie's equation, $V_{sh}$is the volume of shale, and φ_{Ssh} is the reading of sonic log against a shale formation.

It is worth mentioning that, compared to the neutron and density logs, the sonic log is less affected by borehole washouts and hence it can be considered an effective log for porosity calculation in bad hole intervals of tight sandstones.

Density or gamma-gamma log (RHOB) measures the formation density through gamma-ray sources and detectors. Density log reading is inversely correlated to the formation porosity; the lower the density log reading, the lower the porosity. Compared to the neutron and sonic logs, gas presence in the formation has little effect on a density log reading, but when combined with density or sonic logs provides an instrumental tool for gas detection. The density log is too sensitive for washout intervals and hence care must be taken in the porosity calculation of bad hole zones. A sample of the density log is shown in Fig. 7–7. Porosity is calculated from the density log (φ_D) using the following equation.

$$\varphi_D = \frac{\rho_{\text{ma}} - \rho_{\text{log}}}{\rho_{\text{ma}} - \rho_f} \tag{7.16}$$

where ρ_{log} is density log reading at each depth point, $\rho_{\text{ma}}$is matrix density and ρ_f fluid density (1.1 g/cc for the case of brine). ρ_{ma} is taken as 2.55, 2.65, 2.71, and 2.87 g/cc for shale, sandstone, limestone, and dolomite, respectively.

The shale-corrected density porosity (φ_{Dshc}) is calculated using Eq. (7.17).

$$\varphi_{\text{Dshc}} = \varphi_D - (V_{\text{sh}} \times \varphi_{\text{Dsh}}) \tag{7.17}$$

Where φ_D is apparent density porosity, $V_{sh}$is the volume of shale, and φ_{Dsh} is the reading of density log against a shale formation.

Composite logs provide a more robust way for gas detection compared to sing single porosity logs evaluation. Composite neutron-density (ND), neutron-sonic (NS), and density-sonic (DS) logs are used as well log evaluation techniques for accurate porosity calculation. For the case of gas reservoirs, the overlay of neutron and density logs in the same track provides a robust approach for gas detection. The responses of neutron and density log against three lithologies including sandstone, limestone, and dolomite in the presence and absence of gas are shown in Fig. 7–8. As

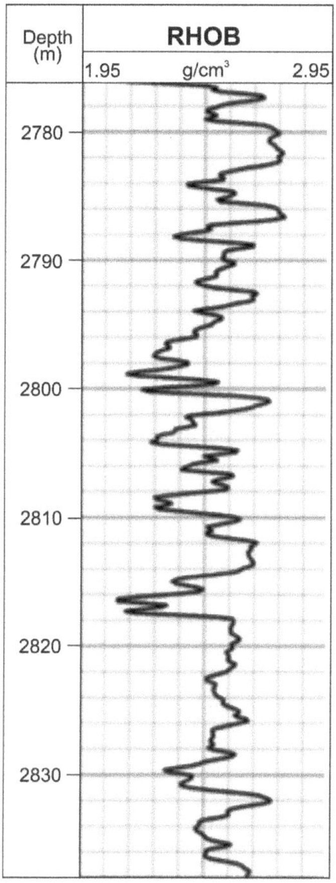

Figure 7–7 A sample of density log. The deviation of the density log to the right indicates dense formation and its deviation to the left indicates porous formation. *Originally generated for this chapter.*

shown in Fig. 7–8A, for the case of oil formations the following neutron and density responses are observed.

- Sandstone: neutron log is to the right side of density log
- Limestone: neutron and density logs coincide,
- Dolomite: neutron log is to the left side of density log

While in a gas-saturated rock (Fig. 7–8B), due to the lower concentration of hydrogen atoms (also called the gas effect (neutron log shifts towards the right side of the track or lower values. Accordingly, the gas effect alters neutron and density responses as follows.

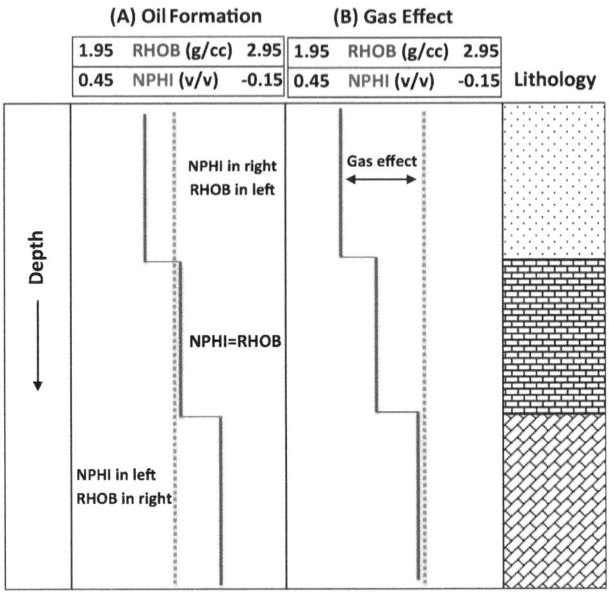

Figure 7–8 The response of neutron and density logs against sandstone, limestone, and dolomite in case of oil formations (A) and gas formations (B). The gas effect causes the shift of neutron log to the right limit of the plots leading to a higher neutron-density separation in sandstones. *Originally generated for this chapter.*

- Sandstone: neutron log shifts to the right of density log causing a higher separation between them. Such a high ND separation or crossover occurs in gas-bearing sandstones.
- Limestone: neutron shifts to the right of density log. Such an overlay may wrongly be interpreted as sandstone, while it is due to the gas effect causing the neutron log shifts to the right of the density log.
- Dolomite: rightwards shift of neutron log may cause the neutron log to coincide with density log as with the limestone.

7.2.2.1 Porosity overestimation/underestimation using the conventional methods

Tight gas sandstones are heterogeneous with ultra-low porosity and permeability and a complex pore system. Using the conventional methods of petrophysical evaluation will lead to a lot of uncertainties in rock properties estimation. The total porosity of tight gas sandstones is a little greater than the core-derived porosity and the effective porosity derived from the well logs analysis is close to or lower than the core-derived porosity. The presence of clay minerals intensifies this effect such that in the case of high clay volumes, the well log-derived total porosity is much higher than the core porosity and similarly the well log-derived effective porosity exceeds the core porosity (Moore et al., 2016). The source of clay minerals is mainly attributed to the alteration and decomposition of feldspars into illite, chlorite, and other swelling clays. Porosity is usually estimated from neutron, density, and sonic logs. The log-derived porosity is compared to the core-derived porosity to have an acceptable estimation of reservoir porosity. However, in the case of tight sands finding such a relationship seems to be difficult due to high reservoir heterogeneity. For example, in the central Sichuan basin of China, the correlation between porosity logs (neutron, density, and sonic) and core porosity is weak (Fig. 7–9). Although sonic logs show to some extent stronger relationships with core porosity, the porosity of tight gas sandstones needs to be estimated more accurately to have a more realistic reserve calculation. The presence of heavy minerals such as siderite and pyrite in either matrix or cement content affects the well log-derived porosity complicating the relationship between core and well log porosities. The tightest zones or cleanest intervals are not always the best reservoir units due to the effect of diagenetic overprints. The logged intervals of tight gas sandstones suffer from bad hole conditions, rogues hole, and washouts and hence it is important to use bad hole models to prevent errors in well log data analysis. It should be noted that due to the high formation pressure in tight gas reservoirs and their low invasion profile caused by low permeability it is very important to perform environmental correction and gas effect before porosity calculation.

7.2.2.2 Porosity calculation from the combination of sonic and nuclear magnetic resonance log

As mentioned earlier, due to the complex pore structure of tights sands using conventional methods for porosity calculation is associated with high uncertainties. For this reason, the application of advanced logging techniques such as NMR log is an effective way to understand tight sands pore structure. In water-saturated or oil-bearing formations, the area under the NMR T2 distribution curve is an excellent estimation of porosity, while in a gas reservoir the extracted NMR porosity is underestimated due to the low hydrogen index of gas. For this reason, a combination of porosity derived from NMR log and conventional porosity logs is a good solution to have an accurate estimation of porosity in tight gas sandstones. Neutron and density logs measure the formation of total porosity, while the sonic log response is an indication of connected or effective porosities. Accordingly, an effective way is a combination of the NMR porosity (also calledφ_{NMR}) with the sonic log porosity.

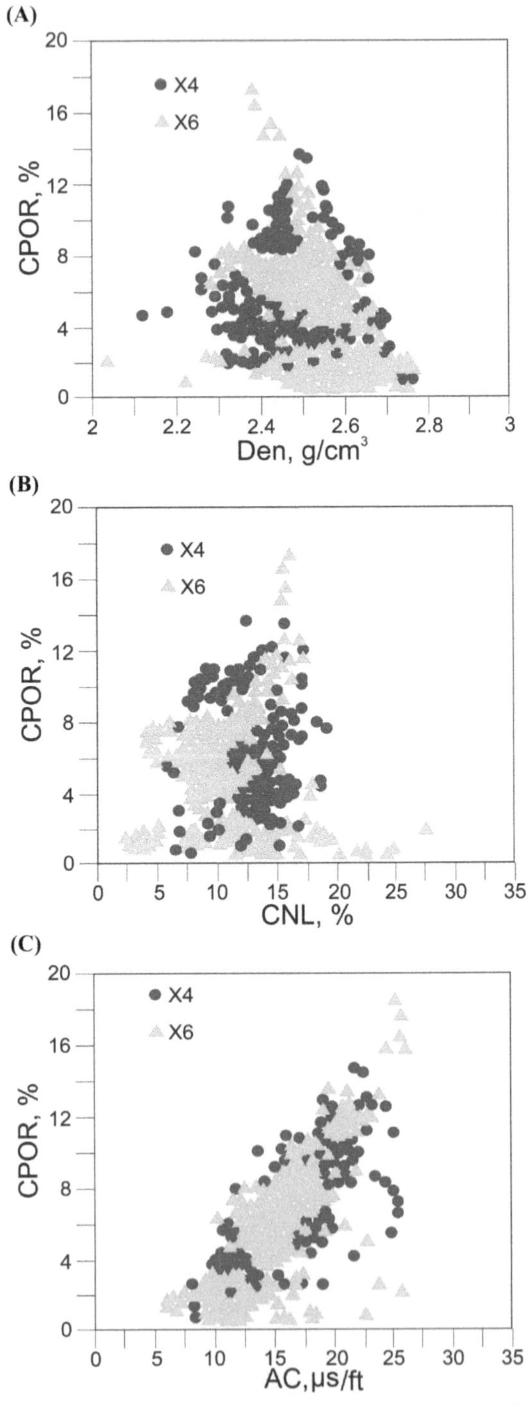

Figure 7–9 The relationship between core-derived porosity (CORP) and porosity logs including density (A), neutron (B), and sonic (C) logs in the Sichuan basin, China (Xiao et al., 2015).

Coates et al. (2000) and Dunn et al. (2002) proposed the following equation for the correction of NMR porosity.

$$\varphi_{\mathrm{NMR}} = \varphi \times S_g \times HI_g \times P_g + \varphi \times HI_f \times (1 - S_g) \tag{7.18}$$

where polarization factor P_g is calculated as follows.

$$P_g = 1 - e^{-\frac{T_W}{T1_g}} \tag{7.19}$$

In the case of fully water-saturated rocks, Eq. (7.18) can be simplified as Eq. (7.20).

$$\frac{\varphi_{\mathrm{NMR}}}{\varphi} = 1 - S_g \times \left(1 - HI_g \times P_g\right) \tag{7.20}$$

As mentioned, the general form of Wyllie's equation for porosity calculation from the sonic log is expressed as follows.

$$\varphi_S = \frac{\mathrm{DT}_{\log} - \mathrm{DT}_{\mathrm{ma}}}{\mathrm{DT}_{\mathrm{f}} - \mathrm{DT}_{\mathrm{ma}}} \tag{7.21}$$

Wyllie's average time equation for tight gas sandstones can be written as follows.

$$\mathrm{DT}_{\log} = \mathrm{DT}_{\mathrm{ma}} \times (1 - \varphi) + \mathrm{DT}_w \times \varphi \times (1 - S_g) + \mathrm{DT}_g \times \varphi \times (S_g) \tag{7.22}$$

Replacing Eq. (7.22) into Eq. (7.21), the following equation can be obtained:

$$\frac{\varphi_S}{\varphi} = 1 + S_g \times \left(\frac{\mathrm{DT}_{\log} - \mathrm{DT}_{\mathrm{ma}}}{\mathrm{DT}_f - \mathrm{DT}_{\mathrm{ma}}}\right) \tag{7.23}$$

As is seen from Eq. (7.23), calculating porosity from the sonic log is highly dependent on S_g and DT_f (Xiao et al., 2012, 2015) which is difficult to be achieved based on only conventional transit time. That is why incorporating NMR results into the porosity equation will aid in more precise porosity estimation. The two parameters α β can be introduced as follows.

$$\alpha = \frac{\mathrm{DT}_{\log} - \mathrm{DT}_{\mathrm{ma}}}{\mathrm{DT}_f - \mathrm{DT}_{\mathrm{ma}}} \tag{7.24}$$

$$\beta = 1 - \mathrm{HI}_g \times P_g \tag{7.25}$$

Having α and β parameters substituted into Eqs. (7.18) and (7.24) and merging them, the following equation can be obtained.

$$\varphi = \left(\frac{\beta}{\alpha+\beta}\right) \times \varphi_S + \left(\frac{\alpha}{\alpha+\beta}\right) \times \varphi_{NMR} \tag{7.26}$$

By further definition of $C_1 = \frac{\beta}{\alpha+\beta}$ and $C_2 = \frac{\alpha}{\alpha+\beta}$, Eq. (7.27) can be simplified as follows.

$$\varphi = C_1 \times \varphi_S + C_2 \times \varphi_{NMR} \tag{7.27}$$

where

$$C_1 + C_2 = \frac{\beta}{\alpha+\beta} + \frac{\alpha}{\alpha+\beta} = 1 \tag{7.28}$$

The nomenclature for the terms used in the Eqs. (7.18) through (7.28) is as follows.

φ_{NMR}: NMR porosity in the fraction
φ: formation porosity (or core porosity) in the fraction
S_g: gas saturation in the fraction
HI_g: gas hydrogen index in the fraction
HI_f: pore fluid hydrogen index in the fraction
P_g: polarization factor
T_w: polarization time in μs
T_{1g}: gas longitudinal relaxation time in μs
DT_{log}: sonic log measurement against formation in μs/ft
DT_{ma}: transit time of matrix in μs/ft (55 μs/ft for sandstone)
DT_f: transit time of pore fluid in μs/ft
DT_w: transit time of water in μs/ft (189.5 μs/ft)
DT_g: transit time of gas in μs/ft
C_1: sonic porosity coefficient
C_2: NMR porosity coefficient

C_1 and C_2 coefficients are calibrated by using core porosity data. Dividing both sides of Eq. (7.27) byφ_{NMR}, Eq. (7.29) is derived as follows.

$$\frac{\varphi_{Core}}{\varphi_{NMR}} = C_1 \times \frac{\varphi_S}{\varphi_{NMR}} + C_2 \tag{7.29}$$

Comparing Eq. (7.29) with the simple regression equation $y = ax + b$, C_1 C_2 coefficients can be obtained through plotting $y = \frac{\varphi_{Core}}{\varphi_{NMR}}$ versus $x = \frac{\varphi_S}{\varphi_{NMR}}$.

An example of calibrating C_1 and C_2 coefficients in the Xujiahe Formation of the central Sichuan basin is shown in Fig. 7–10. As is seen, the final estimation of tight sands porosity is achieved through the following equation.

$$\varphi = 0.809 \times \varphi_S + 0.191 \times \varphi_{NMR} R^2 = 0.89 \tag{7.30}$$

Prior to using Eq. (7.30) for porosity calculation in tight sands, sonic porosity should be corrected for shale effect and gas effect.

7.2.3 Water saturation

One of the main challenges associated with the tight gas sandstones is their low resistivity behavior making it difficult to be discriminated from fully water-saturated formations. The main pitfalls in this regard can be mentioned as low water saturation, undetectable or very small transition zone, reservoir discontinuity, abnormal pressure, and difficulty in free water level identification. Among the main challenges is an underestimation of water saturation which is sometimes called "sub-irreducible

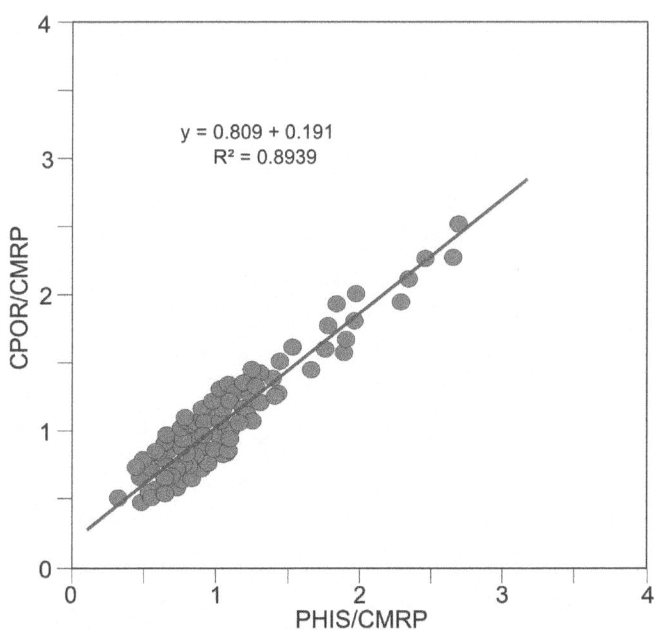

Figure 7–10 Calibration of C_1 and C_2 coefficient in tight sandstones of the central Sichuan Basin, China (Xiao et al., 2015).

water saturation." Some other researcher (Crotti, 2007) believes that water saturation is nearly equal to irreducible water saturation or rarely exceeds that.

That is why they are sometimes treated as low resistivity/low contrast (LRLC) reservoirs. For example, the deep resistivity reading in a gas-producing tight sandstone in the central Sichuan basin is about 9 Ω.m, while the nearby water-producing zone shows the LLD reading of about 5 Ω.m. Both zones have a similar density reading showing similar porosities and resistivities, but the DST results indicate the successful gas flow in the first zone.

Generally, the historic Archie's (1942) equation is used for water saturation calculation in clean sandstone formations. The mathematical expression of Archie's equation is as follows.

$$S_w = \sqrt[n]{\frac{a.R_w}{\phi^m.R_t}} = \left(\frac{a.R_w}{\phi^m.R_t}\right)^{1/n} \tag{7.31}$$

where S_w is water saturation, ϕ is porosity, R_w is formation water resistivity, R_t is true formation resistivity. The parameters m, n and a are called Archie's coefficients. The parameter a is the tortuosity factor, m and n are the cementation factor and saturation exponent, respectively.

Usually, the reading of the deep resistivity log (such as LLD, ILD, or RLA5) after environmental corrections is taken as the uninvaded zone resistivity (R_t). R_w is calculated based on formation water salinity test results and Schlumberger charts. In the absence of formation water analysis results, the standard petrophysical charts (e.g., Pickett plot or Hingle plot) or other theoretical models are used to estimate R_w.

Determination of Archie's coefficient has an important role in the quantification of tight gas water saturation. In order to derive m and parameters, the ratio of R_o to R_w or (formation factor, F) is measured on core plug samples. Each core plug is saturated with formation brine and its corresponding rock resistivity (R_o) and water resistivity (R_w) are measured to derive the formation factor using Eq. (7.32).

$$F = \frac{R_o}{R_w} \tag{7.32}$$

A log-log plot of formation factor versus porosity is prepared for all core plug samples and a regression line is fitted to all samples. The slope and intercept of the fitting line represent the cementation factor and tortuosity coefficient, respectively. An alternative method to obtain F is using Eq. (7.33).

$$F = \frac{a}{\phi^m} \tag{7.33}$$

Similarly, a cross-plot of resistivity index ($RI = R_t/R_o$) versus water saturation is prepared for the analyzed core plugs, and the degree of fitting line's curvature is taken as the saturation exponent.

In the tight gas sandstones, we need more accuracy to determine water saturation which is highly dependent on Archie's parameters and formation water resistivity. For example, the relationship between resistivity index and water saturation in the F4 and F6 Formation of the Sichuan basin is so scattered that choosing an appropriate saturation exponent is difficult (Fig. 7–11). If the relationship between *Log F* versus *Log* ϕ or *RI* versus S_w in tight gas reservoirs is scattered and complex, then Archie's equation will not produce an accurate estimation of water saturation. Considering the critical low porosity and low permeability of the tight gas reservoirs, minor errors in water saturation estimation will lead to significant errors in their reserve estimation and economic evaluation. Several case studies around the world have shown that most tight formations are associated with clay minerals and shaly interbeds affecting the

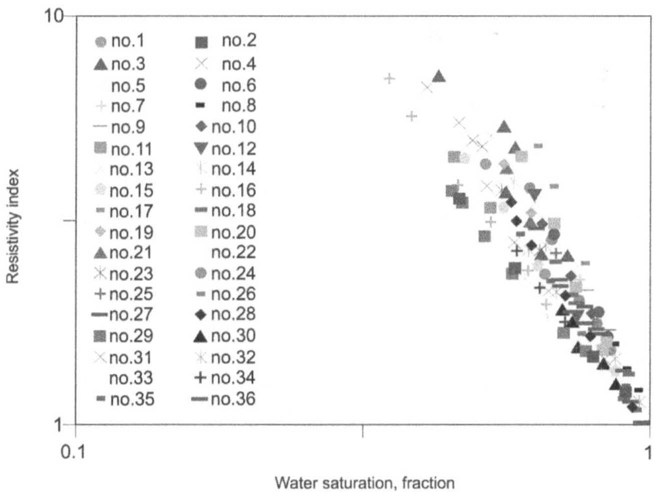

Figure 7–11 Resistivity index versus water saturation cross plot for saturation exponent determination in the tight sandstones of F4 and F6 Formations, the Sichuan basin, southwestern china. Determination of saturation exponent is associated with a high uncertainty due to high data scattering in low saturation (Xiao et al., 2013).

electrical properties of the formation. The presence of clay minerals in tight gas sandstone reservoirs reduces the true resistivity of the formation and leads to an overestimation of water saturation. The negative charges on the surface of two-layered and three-layered clay minerals interact with bipolar cations and water molecules. Considering the low permeability nature of tight sandstones, the presence of clay minerals, even in small amounts, needs to be accounted for to have a realistic gas in place calculation. In this regard, the Archie equation will not be applicable and it is necessary to use the models that account for the shale effect. One of the popular models for water saturation calculation in shaly tight sandstone is the Waxman-Smith equation (1968).

$$C_t = \frac{C_w}{F*} S_w^n + \frac{BQ_v}{F*} S_w^{n-1} \tag{7.34}$$

where C_t is electrical conductivity in the hydrocarbon zone, C_w formation water conductivity, F is formation factor and Q_v is cation exchange capacity per unit volume which (Table 7—1).

$$Q_v = \frac{\text{CEC}*(1 - \varphi_t)\rho_{\text{ma}}}{\varphi_t} \tag{7.35}$$

Parameter B represents the equivalent electrical conductivity of sodium for the exchange ions of clay and is a function of temperature (T) and water resistivity of the formation (R_w) at 25°C which is calculated as follows.

$$B = \frac{-1.28 + (0.225 \times t) - (0.0004059 \times t^2)}{1 + R_w^{1.23} \times (0.045 \times t - 0.27)} \tag{7.36}$$

Sarihi and Vargas-Murillo (2015) introduced a new term, named the clay factor (C_f) to account for the quantity of the exchangeable clay. The term C_f could replace the BQ_v in the Waxman-Smits correlation. The new equation works based on modifications of the Waxman-Smits equation considering conductivity and volume of clay.

Table 7—1 The equivalent Q_v values versus shale volume.

Shale volume	Q_v
High	1.5
Intermediate	1
Low	0.5
Very low	0.25
Shale-free	0

It should be noted in the absence of any laboratory data, all water saturation equations (e.g., Dual Water, Indonesia, Simandoux, etc) shall be tried and compared with the reservoir porosity and lithological data. For example, it is expected that very dense and ultra-tight zones of tight sandstones shall have a higher water saturation in their nano-pores. This way, a quick-look judgment about "which water saturation estimation model performs better" can be made. Overall, considering the low porosity and low permeability nature of the tight gas sandstone, it is vital to calibrate the well log-derived saturation with core and NMR results to avoid errors in their reserve estimation and economic evaluation.

7.3 Probabilistic method

When a complete suite of well logs is available in a tight gas field, the use of probabilistic (Multimin) methods provides satisfactory results. In the probabilistic methods, in addition to the existence of sufficient wells, the petrophysicist must have a good knowledge of the target formation, depositional setting, and regional geology of the study area. Unlike the deterministic method, in which petrophysical calculations are performed step by step, in the probabilistic method, the calculations are performed simultaneously by solving multi-mineral equations. In the probabilistic method, by solving the polynomial equations, the mineral and fluid volumes are calculated. The general form of the probabilistic equations for the case of six unknowns is as follows.

$$\begin{bmatrix} \alpha_{11} & \alpha_{12} & \alpha_{13} & \alpha_{14} & \alpha_{15} & \alpha_{16} \\ \alpha_{21} & \alpha_{22} & \alpha_{23} & \alpha_{24} & \alpha_{25} & \alpha_{26} \\ \alpha_{31} & \alpha_{32} & \alpha_{33} & \alpha_{34} & \alpha_{35} & \alpha_{36} \\ \alpha_{41} & \alpha_{42} & \alpha_{43} & \alpha_{44} & \alpha_{45} & \alpha_{46} \\ \alpha_{51} & \alpha_{52} & \alpha_{53} & \alpha_{54} & \alpha_{55} & \alpha_{56} \\ \alpha_{61} & \alpha_{62} & \alpha_{63} & \alpha_{64} & \alpha_{65} & \alpha_{66} \end{bmatrix} \times \begin{bmatrix} A \\ B \\ C \\ D \\ E \\ F \end{bmatrix} = \begin{bmatrix} \beta_1 \\ \beta_2 \\ \beta_3 \\ \beta_4 \\ \beta_5 \\ \beta_6 \end{bmatrix}$$

where α_{ij} is the standard responses for minerals and fluids, A, B, C, $D...F$ are the mineral and fluid volumes to be estimated, and β_i is the reading of well logs.

The above equations are only generated for a specific depth point of the formation, and this system of equations must be solved for every depth point separately to obtain the minerals and fluid volumes as a continuous log for the entire logged interval.

For example, if we have six unknowns including volumes of oil, free water, quartz, calcite, dolomite, and illite, the measured value or density response of the formation is estimated by Eq. (7.37).

$$\rho_b = \rho_{oil}v_{oil} + \rho_{fw}v_{fw} + \rho_{qtz}v_{qtz} + \rho_{cal}v_{cal} + \rho_{dol}v_{dol} + \rho_{illite}v_{illite} \quad (7.37)$$

where ρ_b is the density log reading. The same equations are generated for other well log responses such as gamma-ray, neutron, acoustic, resistivity, etc to generate at least six equations (for this example) that can be solved.

The weight of each petrophysical log is included in the calculations by assigning an uncertainty value as a coefficient. Well logs with low uncertainty will have a low weight in calculations and vice versa. For example, at a washout interval, the uncertainty of the sonic log should be decreased to increase its weight in porosity calculation. Because the sonic log is not sensitive to the borehole wall washouts. Instead, the uncertainty of the density log should be increased, which is extremely affected by washouts. To optimize the calculations, a set of upper and lower bounds are applied to the mineral and fluid volumes. For example, porosity is set to be calculated between 0 and 0.3, or water saturation is limited in the range of 0 and 1, and the out-of-range values are replaced by the upper or lower limits.

Inverse modeling: When the volumes of minerals and fluids are calculated, the same system of equations is used to obtain the response of the well logs for the calculated volumes. In this way, a series of reconstructed log will be established and the comparison of the actual well logs with the reconstructed ones will indicate the accuracy of the probabilistic model. The closer the actual logs to the reconstructed ones, the higher the accuracy of the probabilistic model. It should be noted that the number of unknowns in the equation should not exceed the number of known variables (i.e., well logs). For example, if four unknowns are to be estimated, including volumes of shale, quartz, clay, and porosity, there must be at least three petrophysical log such that the established equations are solvable. The fourth equation is obtained from the material balance model, which states that the sum of the mineral and fluid volumes for each depth point of the formation equals one.

7.4 Permeability

Permeability estimation is one of the major challenges in rock typing and reservoir characterization of the tight gas sandstone reservoirs. In most conventional sandstone reservoirs, there is a strong correlation between porosity and permeability which can be used for permeability estimating based on the well log-derived porosity. However, such a relationship becomes weak for tight gas reservoirs due to their complex and heterogeneous pore network system. Small amounts of clay minerals can severely affect the low porosity and ultra-low permeability rocks. For example, the scatter-plot of porosity and permeability for X4 and X6 Formations of the Sichuan basin is shown in Fig. 7–12.

As is seen, it is hard to fit a regression line to such scattered data. Hence, using the conventional methods will not be effective in the permeability estimation of tight sandstones. In this regard, a method reaping the benefits of mercury injection capillary pressure (MICP) curves and NMR data will be of prime importance in the high-resolution permeability estimation of tight sandstones.

Using the SDR (Schlumberger Doll Research Center) and Timur–Coates models permeability of the tight sandstones can be estimated from the NMR log as follows (Kenyon, 1997; Kenyon et al., 1988; Coates et al., 1999).

SDR model:

$$k = C_1 \times T_2LM^{a1} \varphi_e^{b1} \tag{7.38}$$

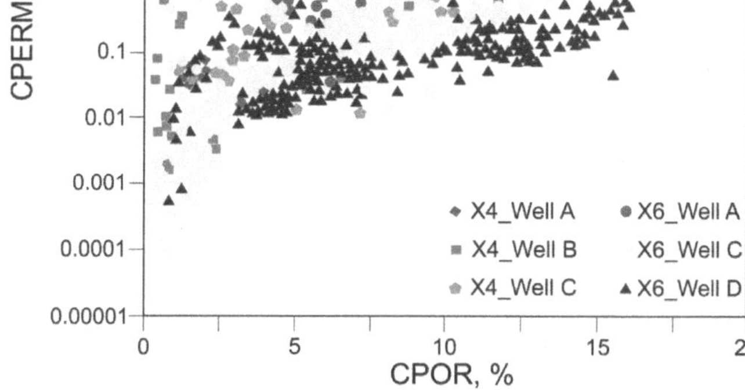

Figure 7–12 The relationship between core porosity and permeability, X4 and X6 Formations of the Sichuan basin (Xiao et al., 2015).

where T_2LM is the logarithmic mean of NMR T_2 distribution and φ_e is effective porosity. a_1, b_1, and C_1 are coefficients that by default are set to 2, 4, and 10.

Timur-Coates model:

$$k = \left(\frac{\mathrm{FFI}}{\mathrm{BVI}}\right)^{a2} \times \left(\frac{\varphi_e}{C_2}\right)^{b2} \tag{7.39}$$

where FFI is the free fluid bulk, BVI is the bulk volume irreducible, and φ_e is effective porosity. a_2, b_2, and C_2 are coefficients that by default are set to 2, 4, and 10.

In order to have a high-resolution permeability estimation based on the NMR log, it is necessary to run NMR experiments on core samples of tight gas sandstone to derive parameters φ_e, T_2LM, FFI, and BVI. Parameters φ_e and T_2LM can easily be derived from NMR log after necessary corrections and processing, while FFI and BVI are dependent on the T_2 cutoffs based on core NMR experiments (Xiao et al., 2014). Afterward, using an optimization algorithm coefficients a_1, b_1, C_1, a_2, b_2 and C_2 are adjusted such that a good agreement between the core-derived permeability and SDR or Timur-Coates permeability is achieved. The obtained coefficients and cutoff are applied for the entire NMR-logged interval to derive high-resolution permeability.

Rezaee et al. (2012) adopted the Winland R35 equation (Kolodzie, 1980) and proposed a new correlation for permeability estimation in the tight gas sandstones of Western Australia. They concluded that the representative pore radius at 35% mercury saturation of conventional reservoirs is not applicable for tight gas sandstones. The pore throat radius corresponding to $S_{Hg} = 35\%$ falls in the range of nano-pores, while R10 can best represent the mean pore radius of tight sandstones falling in the range of 0.1−1.0 μm.

The other popular methods for permeability estimation in tight gas sandstones can be mentioned as the Swanson parameter-based model (Thomeer, 1960; Swanson, 1981; Huet et al., 2007), the Capillary Parachor-based model (Guo et al., 2004), and the modified Winland R35 method (Mao et al., 2013).

References

Archie, G.E., 1942. The electrical resistivity log as an aid in determining some reservoir characteristics. Society of Petroleum Engineers, Transactions of the AIME 146 (1), 54−62.

Clavier, C., Hoyle, W., Meunier, D., 1971. Quantitative Interpretation of Thermal Neutron. Decay Time Logs: Part I. Fundamentals and Techniques. SPE-2658-A-PA.

Coates, G.R., Xiao, L.Z., Prammer, M.G., 1999. NMR Logging: Principles and Applications. Haliburton Energy Services, Houston, TX, p. 234.

Coates, G.R., Xiao, L.Z., Primmer, M.G., 2000. NMR Logging Principles and Applications. Gulf Publishing Company, Houston, TX, pp. 1–256.

Crotti, M.A., 2007. Water saturation in tight gas reservoirs. Society of petroleum engineers. In: Latin American & Caribbean Petroleum Engineering Conference, 15–18 April, Buenos Aires, Argentina. doi:10.2118/107145-MS.

Dunn, K.J., Bergman, D.J., Latorraca, G.A., 2002. Nuclear magnetic resonance: petrophysical and logging applications. Handbook of Geophysical Exploration. Pergamon, New York, pp. 1–176.

Guo, B.Y., Ghalambor, A., Duan, S.K., 2004. Correlation between sandstone permeability and capillary pressure curves. Journal of Petroleum Science and Engineering 43 (3–4), 239–246.

Huet, C.H., Rushing, J.A., Newsham, K.E., Blasingame, T.A., 2007. Estimating Klinkenberg-corrected permeability from mercury-injection capillary pressure data: a new semi-analytical model for tight gas sands. In: Rocky Mountain Oil & Gas Technology Symposium, 16–18 April 2007, Denver, USA, Conference paper 102890-MS.

Kenyon, W.E., 1997. Petrophysical principles of applications of NMR logging. The Log Analyst 38 (3), 21–43.

Kenyon, W.E., Day, P.I., Straley, C., Willemsen, J.F., 1988. A three-part study of NMR longitudinal relaxation properties of water-saturated sandstones. SPE Formation Evaluation 3 (3), 622–636.

Kolodzie Jr., S., 1980. Analysis of pore throat size and use of the Waxman–Smits equation to determine OOIP in spindle field. Colorado, Paper SPE 9382 Presented at the 1980 Annual Fall Technical Conference of Society of Petroleum Engineers. September 21–24.

Larionov, V.V., 1969. Borehole Radiometry (in Russian). Izdat, Nedra, Moscow.

Mao, Z.-Q., Xiao, L., Wang, Z.-N., Jin, Y., Liu, X.-G., Xie, B., 2013. Estimation of permeability by integrating nuclear magnetic resonance (NMR) logs with mercury injection capillary pressure (MICP) data in tight gas sands. Applied Magnetic Resonance 44 (4), 449–468.

Moore, W.R., Zee, Ma, Y., Pirie, I., Zhang, Y., 2016. Tight gas sandstone reservoirs, part 2: petrophysical analysis and reservoir modeling, Chapter 15. In: Zee Ma, Y., Holditch, S.A. (Eds.), Unconventional Oil and Gas Resources Handbook, Evaluation and Development. Gulf Professional Publishing, Oxford, United States, pp. 429–448.

Rezaee, R., Saeedi, A., Clennell, B., 2012. Tight gas sands permeability estimation from mercury injection capillary pressure and nuclear magnetic resonance data. Journal of Petroleum Science and Engineering 88–89, 92–99.

Sarihi, A., Vargas-Murillo, B., 2015. A method to compute water saturation in tight rocks accounting for conductivity of clay minerals. In: Abu Dhabi International Petroleum Exhibition and Conference, 9–12 November, Abu Dhabi, UAE. Society of Petroleum Engineers, SPE-177550-MS.

Serra, O., 1984. Fundamentals of Well-Log Interpretation (Part A): The Acquisition of Logging Data. Elsevier, Amsterdam, p. 422.

Stieber, S., 1970. Pulsed Neutron Capture Log Evaluation – Louisiana Gulf Coast. In: SPE2961-MS, SPE. Society of Petroleum Engineers.

Swanson, B.F., 1981. A simple correlation between permeabilities and mercury capillary pressures. Journal of Petroleum Technology 33 (12), 2498–2504.

Thomeer, J.H.M., 1960. Introduction of a pore geometrical factor defined by the capillary pressure curve. Transactions of the American Institute of Mining, Metallurgical, and Petroleum Engineers 219 (12), 354–358.

Waxman, M.H., Smits, L.J.M., 1968. Electrical conductivities in oil-bearing shaly sands. Society of Petroleum Engineers Journal 8, 107–122.

Xiao, L., Liu, X.-P., Zou, C.-C., Hu, X.-X., Mao, Z.-Q., Shi, Y.-J., et al., 2014. Comparative study of models for predicting permeability from nuclear magnetic resonance (NMR) logs in two chinese tight sandstone reservoirs. Acta Geophysica 62 (1), 116–141.

Xiao, L., Mao, Z.Q., Li, G.R., Jin, Y., 2012. Calculation of porosity from nuclear magnetic resonance and conventional logs in gas-bearing reservoirs. Acta Geophysica 60 (4), 1030–1042.

Xiao, L., Mao, Z.-Q., Yan, J., 2015. Tight gas sandstone reservoirs evaluation from nuclear magnetic resonance (NMR) logs: case studies. Arabian Journal for Science and Engineering 40 (4), 1223–1237.

Xiao, L., Zou, C.C., Mao, Z.Q., Shi, Y.J., Liu, X.P., Jin, Y., et al., 2013. Estimation of water saturation from nuclear magnetic resonance (NMR) and conventional logs in low permeability sandstone reservoirs. Journal of Petroleum Science and Engineering 108, 40–51.

CHAPTER 8

Artificial intelligence in tight gas reservoirs characterization

8.1 Introduction

The application of machine learning (ML) as a type of artificial intelligence (AI) technique, is increasing in reservoir characterization of tight gas sandstones. ML methods such as neural networks (NNs), fuzzy inference systems (FISs), and hybrid neuro-fuzzy systems have widely been used in the estimation of reservoir rock properties such as porosity, the volume of shale, water saturation, permeability, flow units, and elastic rock properties from good logs and seismic data. ML techniques enable geoscientists and engineers to make intelligent formulations between a set of input data (e.g., well logs, seismic attributes) and output data such as reservoir porosity and permeability. Considering the high sensitivity of calculation or estimation of tight gas reservoirs rock properties, it is of prime importance to reduce uncertainty in reservoir geological and dynamic models. In this chapter, the application of different AI techniques in tight gas reservoirs characterization is described with applied examples. AI aids in reducing uncertainty in tight gas rock properties estimation and delineation of pay zones across a gas field.

8.2 Machine learning methods

ML is a data analysis method based on data science or AI technique that helps in filling the information gap, decision making, pattern recognition, and estimating missing data based on the available information. Geoscientists and engineers use ML methods as a complementary processing and interpretation technique for geological data analysis, petrophysical interpretations, seismic reservoir characterization, static reservoir modeling, and dynamic simulation. This way uncertainty in static and dynamic models of a tight gas reservoir will be reduced and more successful field development plans can be achieved. ML models are classified into supervised and unsupervised techniques. Supervised ML models make a formulation between a set of input and output data, while unsupervised models classify

Reservoir Characterization of Tight Gas Sandstones
DOI: https://doi.org/10.1016/B978-0-323-90180-2.00001-3

the available data into a set of groups called clusters. To date, many authors have employed supervised and unsupervised ML methods for reservoir parameters estimation and characterization which can be mentioned Rezaee et al. (2007), Kadkhodaie et al. (2009a,b), Golsanami et al. (2014), Jafari et al. (2014), Golsanami et al. (2015), Nouri-Taleghani et al. (2015), Zahmatkesh et al. (2017), Safaei-Farouji & Kadkhodaie (2022). In the following section, a brief description of the popular ML methods is provided.

8.2.1 Neural networks

NNs mimic the natural behavior of the human brain in memorization, learning, and recall. They have widely been used in rock and fluid properties characterization. Hitherto, various types of NNs have been developed and applied in the petroleum industry from which can be mentioned feed-forward neural networks (FFNNs), probabilistic neural nets (PNNs), Hopfield network, radial basis function (RBF) NNs, Generalized regression neural network (GRNN) and support vector machine (SVM). Although all types of NNs can be employed in reservoir characterization of tight gas sandstones, FFNNs are popular due to their effectiveness in computational time and the accuracy of estimations. Multilayer perceptron (MLP) is the common type of FFNN that uses a back-propagation algorithm to optimize the weights and bias values. A schematic diagram of an MLP structure is shown in Fig. 8–1A. As is seen

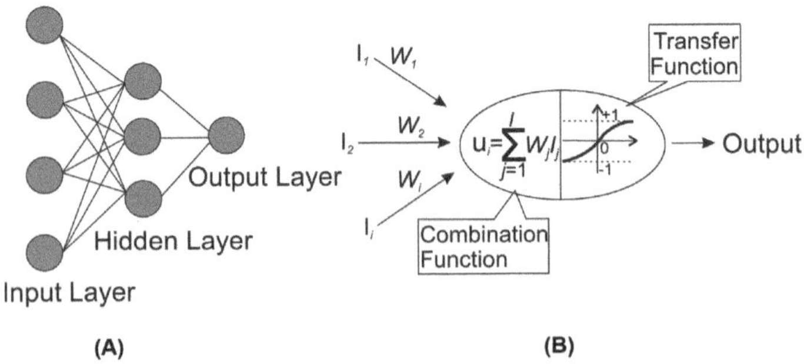

Figure 8–1 Architecture of a multilayer perceptron consisting of input layer, hidden (inner) layer, and output layer (A). A mathematical model of perceptron consisting of a combination function and transfer function (B). The combination function calculates the sum of all inputs multiplied by their weights. When the calculated sum meets the threshold, the transfer function is activated to generate the output.

in Fig. 8–1B, the body of a perceptron consists of two parts including a combination function and an activation function.

The combination function (u_i) sums up all inputs multiplied by their weight coefficients.

$$u_i = \sum_{j=1}^{n} W_j I_j + b \tag{8.1}$$

where I denote the inputs and W is their corresponding weights. The constant b is the bias value that acts as an adjuster. Bias is an additional input shifting the transfer function through increasing or decreasing the combination function. Its role in MLP is analogous to the role of constant in a regression line. When the calculated sum meets a threshold, the transfer function (also called the activation function) is activated to generate the output. The most common activation functions are linear, logarithmic sigmoid, and hyperbolic tangent (Zeidenberg, 1990; Bishop, 1995; Bhatt & Helle, 1999).

Linear:

$$f(n) = n \tag{8.2}$$

Logarithmic sigmoid:

$$f(n) = \frac{1}{1 + e^{-cn}} \tag{8.3}$$

Hyperbolic tangent:

$$f(n) = \frac{e^n - e^{-n}}{e^n + e^{-n}} \tag{8.4}$$

Deep learning is an advanced and complex type of ML method with multiple layers that uses an extensive dataset along with complex algorithms to create an intelligent model. In deep learning, a convolutional neural network (CNN) is applied to formulate input to output data. A convolutional NN is a regularized version of a MLP with multiple layers between the input and output layers (Krizhevsky et al., 2012; Hinton et al., 2012).

Input data for ML models are frequently and easily available in all wells across a hydrocarbon field such as conventional good logs (e.g., neutron, sonic, density, gamma-ray, and resistivity logs), seismic attributes or measurements while drilling (MWD) drill sensor data. Tight gas rock properties such as porosity, permeability, pore size distribution, and elastic properties are the main outputs used in reservoir characterization.

8.2.2 Fuzzy inference system

A FIS or fuzzy logic is an alternative method of solving estimation problems in the petroleum industry. Fuzzy logic tries to quantify the uncertainty associated with estimation problems. Statistical methods such as regression analysis try to minimize the estimation error, while fuzzy systems use this error or uncertainty as an effective tool to improve estimations. A FIS is composed of a set of rules formulating the input to output data.

In this method, first, the input and output space should be partitioned by using a clustering algorithm. Fuzzy c-means clustering and subtractive clustering are the popular methods to partition input and output data (Chen, & Wang, 1999; Chiu, 1994). A fuzzy membership function (e.g., Gaussian membership function) is assigned to each cluster. For example, the mean (μ) and standard deviation (σ) of data within each cluster are used to generate Gaussian membership as follows.

$$F(x) = \frac{1}{\sqrt{2\pi\sigma}} e^{-\left(\frac{x-\mu}{\sigma}\right)^2} \tag{8.5}$$

The main input of the subtractive clustering is the clustering radius (r) which ranges from 0 to 1. A small clustering radius means more small clusters, while a large clustering radius results in fewer large clusters. Steps to using a FIS (Sugeno-type) to estimate sonic logs from neutron and density logs are graphically illustrated in Fig. 8–2. As is seen, the input and output space is divided into three clusters resulting in three *if-then* fuzzy rules. A Gaussian membership (MF) function is assigned to each cluster. Accordingly, three Gaussian distributions are fitted to the low, medium, and high values of the neutron, density, and sonic logs. There are four processing steps to apply a FIS for estimation problems as follows (Fig. 8–2).

Step 1. Fuzzify inputs: The crisp inputs are exposed to the fuzzy membership functions to derive the grade to which the input data belong to each MF.

Step 2. Apply fuzzy operator and truncation method: If the antecedent of a fuzzy rule includes more than one part, then the fuzzy operators (such as AND, OR, NOT, etc) are employed to derive a single rule representing the antecedent result of that rule. The output MF is then truncated by the value determined from the result of applying a fuzzy operator on the antecedent. For this case, AND operator is employed,

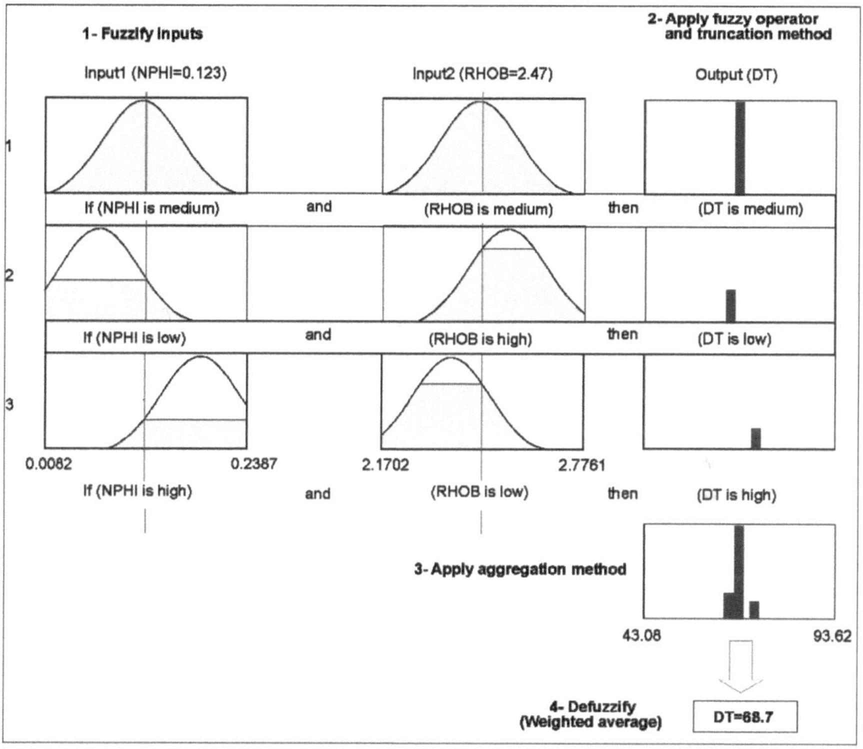

Figure 8–2 Graphical illustration showing the steps to use a fuzzy inference system to estimate sonic log (DT) from neutron (NPHI) and density (RHOB) logs (Rezaee et al., 2008).

Step 3. Apply aggregation method: The results of truncation are combined to form a single fuzzy set.

Step 4. Defuzzify: The result of aggregation which is a fuzzy set is defuzzified using an appropriate method (such as weighted averaging) to obtain a crisp value.

8.2.3 Neuro-fuzzy

Hybrid NN-fuzzy systems, also called adaptive neuro-fuzzy inference systems (ANFIS), reap the advantages of both fuzzy systems dealing with explicit knowledge and NNs dealing with implicit knowledge. In a hybrid neuro-fuzzy model, membership functions are derived based on fuzzy logic, and their parameters (e.g., mean and standard deviation for Gaussian function) are optimized by using NNs (Nikravesh and Aminzadeh, 2003). A five-layered neuro-fuzzy system using multiple

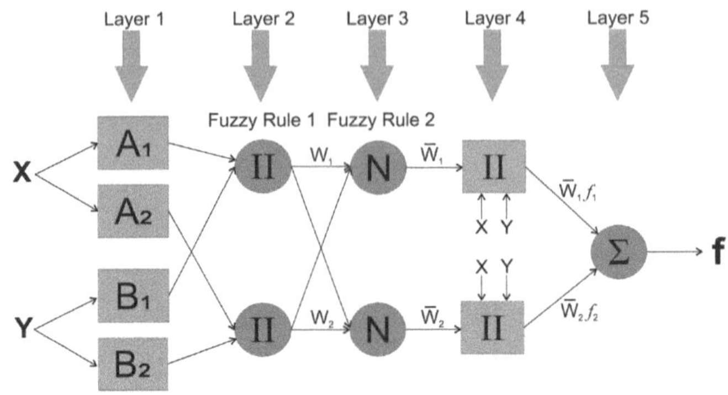

Figure 8–3 A schematic structure of a neuro-fuzzy system (Wang et al., 2003).

fuzzy rules is shown in Fig. 8–3 (Kamali and Mirshady, 2004). As is seen in Fig. 8–3, the architecture of an ANFIS model is composed of five layers as follows. Layer 1 receives the inputs and clusters them into fuzzy sets. Layer 2 generates a fuzzy rule-base based on fitting membership functions to the fuzzy clusters. Layer 3 tunes the membership function parameters based on a back-propagation NN. Layers 4 evaluates the fuzzy rules and generates a combined fuzzy set. Layer 5 outputs a real number based on an appropriate defuzzification method.

8.2.4 Ensemble machine learning model

Ensemble ML model also called committee machine or a committee of experts consists of a group of individual ML models that combines the outputs of each model and thus reaps the advantages of all individual experts, with some additional computations. A committee machine has a parallel structure that generates a final output by combining the outputs of individual ML models by using an optimization method (Haykin, 1999; Sharkey, 1996; Kadkhodaie et al., 2009a,b). The experts may be NNs, fuzzy systems, or other intelligent models (Chen and Lin, 2006). The advantage of this approach is that it can lead to considerable accuracy in the performance of estimations. A schematic diagram of a committee machine is shown in Fig. 8–4. Committee machines are classified into static and dynamic types (Haykin, 1999). In the static committee machines, the outputs of individual experts are combined by an optimization method that does not involve the input signal, hence the structure is static. Two common examples of the static structures are ensemble averaging, where the individual expert's outputs are linearly combined, and

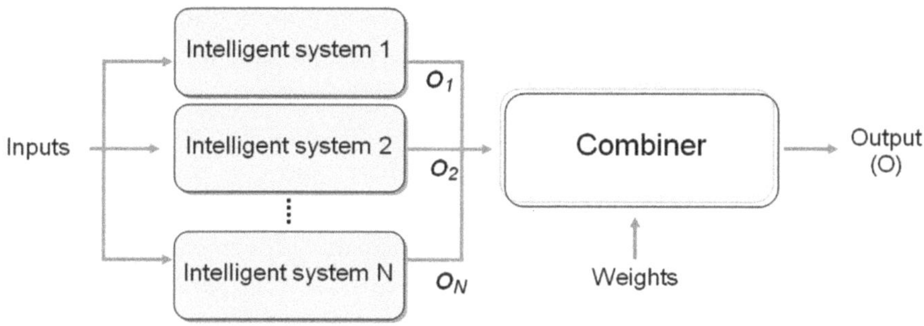

Figure 8–4 A schematic diagram of a committee machine that combines different machine learning models in a combiner by using an optimization method such as a genetic algorithm or ant colony optimization (Kadkhodaie et al., 2009a).

Boosting, where the weak learners are combined to give a strong learner. In the dynamic committee machines, the input signal is directly involved in a combination of outputs and hence they are called dynamic. A good example is Mixtures of experts, where the expert's outputs are non-linearly combined by some form of Gating System than can itself be a NN (Bullinaria, 2008).

8.3 Artificial intelligence in geological reservoir characterization

Geological characterization of tight gas sandstones is of prime importance in understanding the reservoir quality and its interaction with rock and fluid flow heterogeneities. AI plays a pivotal role in the estimation of reservoir properties such as mineralogy, lithofacies, rock types, pore types, faults, and fractures. Some basic concepts of facies analysis used in tight gas sandstones description are explained in brief as follows.

Lithofacies and rock types: *Lithofacies* is a general term used to classify rocks having some sedimentological properties in common under the microscope including mineralogy, texture, fossil content, digenetic history, and depositional environment. Lithofacies can either be classified as microfacies or petrofacies. The term microfacies is used when we deal with carbonate rock and is referred to as petrofacies when we study clastic rocks (e.g., tight sandstones). There are several classification schemes for lithofacies description. Pettijohn et al. (1987) and Folks' classifications (1959, 1962) are the commonly used classification schemes to group petrofacies in tight sandstones. Identification and prediction of *diagenetic facies* based on the petrographic studies, XRD (X-ray diffraction), SEM (scanning electron microscopy), and intelligent systems have an effective contribution toward a more realistic reservoir static model. Diagenetic overprints such as dissolution and fracturing enhance the reservoir quality, while compaction, stylolitization, silica, calcite, and clay minerals cementation have a destructive role in tight reservoirs quality.

Storage capacity (porosity) and flow capacity (permeability) are the main factors considered for the identification of *rock types* and *flow units* in tight gas sandstones. Rock typing of tight gas sandstones is associated with complexities as their pore system is composed of numerous nanopores causing strong non-linear gas flow. This may cause some uncertainties in using conventional rock typing indices for tight gas reservoirs.

AI as a fast, robust and cost-effective computational tool has widely been used to characterize and estimate the lithofacies and rock types of tight gas sandstone reservoirs. Some recent examples utilizing advanced ML methods are outlined as follows. Lu et al. (2020) classified the tight sandstones of the Chang-8 reservoir in the Ordos Basin of China into five diagenetic facies including "chlorite cemented facies, moderately-compacted facies, moderately-dissolved kaolinite-bearing facies, moderately-compacted carbonate cemented facies and strongly-compacted facies." Afterward, they applied a supervised self-organizing map (SSOM) to estimate the diagenetic facies of the tight sandstone from wireline logs and observed a good correlation of 83.87% with core data.

Ji et al. (2022) introduced three indices Fi (Darcy index), Kc (gas-non Darcy index), and Sr (the liquid non-Darcy index) for rock typing of tight sandstones with an example from the Yanchang Formation in Ordos Basin. They used an intelligent model based on a fuzzy k-nearest neighbor (fuzzy KNN) algorithm to optimize the rock typing through the optimal combination of all indices.

Liu and Liu (2022) applied deep learning NNs for lithofacies prediction from well log data including gamma-ray (GR), spontaneous potential (SP), acoustic (AC), and deep resistivity (R_{LLD}). For this purpose, they reaped the advantages of convolutional NNs (CNN) integrated with the long short-term memory networks (LSTM) for lithofacies classification based on well logs. The hybrid CNN-LSTM outperformed the individual deep learning models and provided a more realistic estimation of core-derived facies (Fig. 8—5).

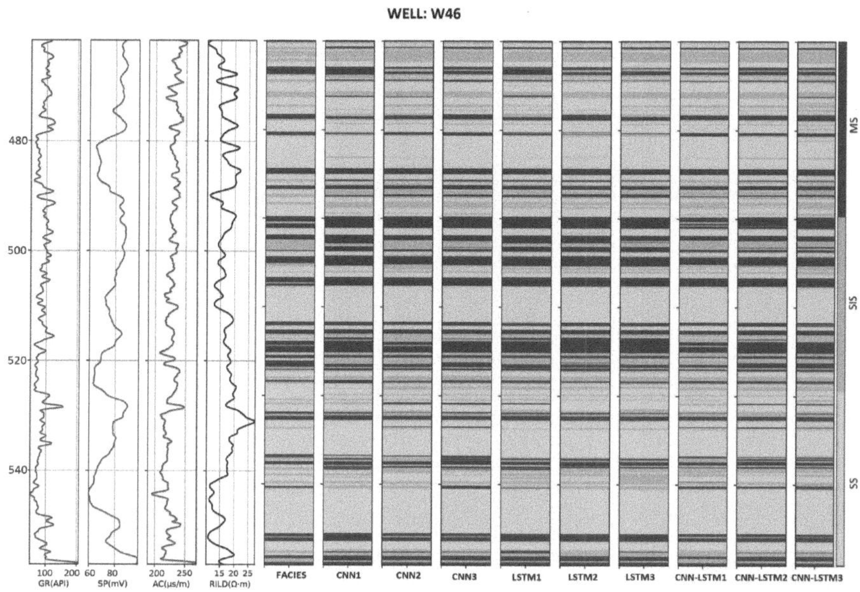

Figure 8–5 The results of lithofacies classification using various deep learning neural networks. The hybrid convolutional neural network-long short-term memory network outperforms the individual convolutional neural networks and long short-term memory networks compared to the core-derived facies (Liu and Liu, 2022).

Fracture characterization: Natural fractures prediction plays a pivotal role in reservoir characterization and modeling of tight gas sandstones. They can easily be identified in the outcrops based on regional geology and tectonic studies, while in subsurface studies seismic data and image logs are the main source of fractures interpretation. There are different tools for acquiring image logs such as Fullbore Formation Microimager (FMI), Simultaneous Time and Resistivity (STAR), Electrical Micro-Imager (EMI), Resistivity-At-Bit (RAB), UBI Ultrasonic Borehole Imager (UBI), Circumferential Acoustic Scanning Tool (CAST), Borehole Televiewer (BHTV) and Circumferential Borehole Image Log Tool (CBIL). FMI, STAR, EMI, and RAB tools provide resistivity-based images, while UBI, CAST, BHTV, and CBIL work based on acoustic contrast.

Different types of fractures such as open fractures (major, medium, minor, and hairline), closed fractures, and continuous, discontinuous, and partially filled fractures can be interpreted based on image logs interpretation. A sample of high angle fractures and representative core samples of

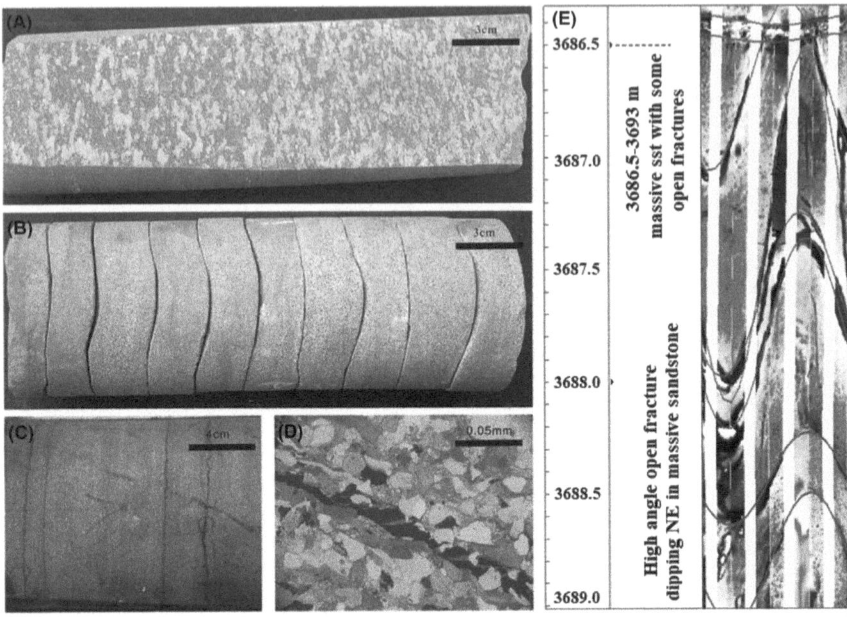

Figure 8–6 Representative core samples showing some high-angle open fractures (A), horizontal fractures (B), diagenetic fractures parallel to bedding (C), fractures related to overpressures and an image log showing high-angle open fractures (E), the Upper Triassic Xujiahe formation, Western Sichuan Basin, China (Gong et al., 2016).

open and close fractures from the tight sandstones of the Ordos Basin (China) are shown in Fig. 8–6.

Moreover, fracture parameters such as density, intensity, porosity, and permeability together with fracture azimuth and dip are additional information derived from image logs analysis.

Using conventional good log data as an indirect source of fractures identification is an efficient and cost-effective method for tight gas reservoirs characterization. However, weak responses of well logs to fractures or difficulty in finding a direct fracture indicator well log imposes a limitation in fractures identification from well logs. In this regard, statistical and intelligent systems such as NNs, and FISs boosting serve as valuable tools for enhancing well logs responses to fractures and formulating conventional well logs to fracture parameters derived from image logs interpretation results. Many studies have shown that a set of conventional well logs such as neutron, density, sonic, photoelectric, and deep resistivity logs are suitable inputs for fracture density estimation (Tokhmchi et al., 2010; Ja'Fari et al., 2011).

Dong et al. (2020a) introduced a fracture indicating parameter (FIP) approach for fracture identification from well logs. For this purpose, they employed a comprehensive index method (CIM) to amplify fracture responses and a comprehensive fractal method (CFM) to characterize fractures based on the fractal approach. Dong et al. (2020b) in a similar study from the tight sandstones of the Ordos Basin applied semi-supervised training algorithms called Laplacian support vector machine (LapSVM) and fine Gaussian SVM that combine Kernel and clustering methods for fractures identification from well-logging data (Fig. 8–7). LapSVM outperforms the fine Gaussian SVM method for the prediction of core-seen fractures based on well log responses.

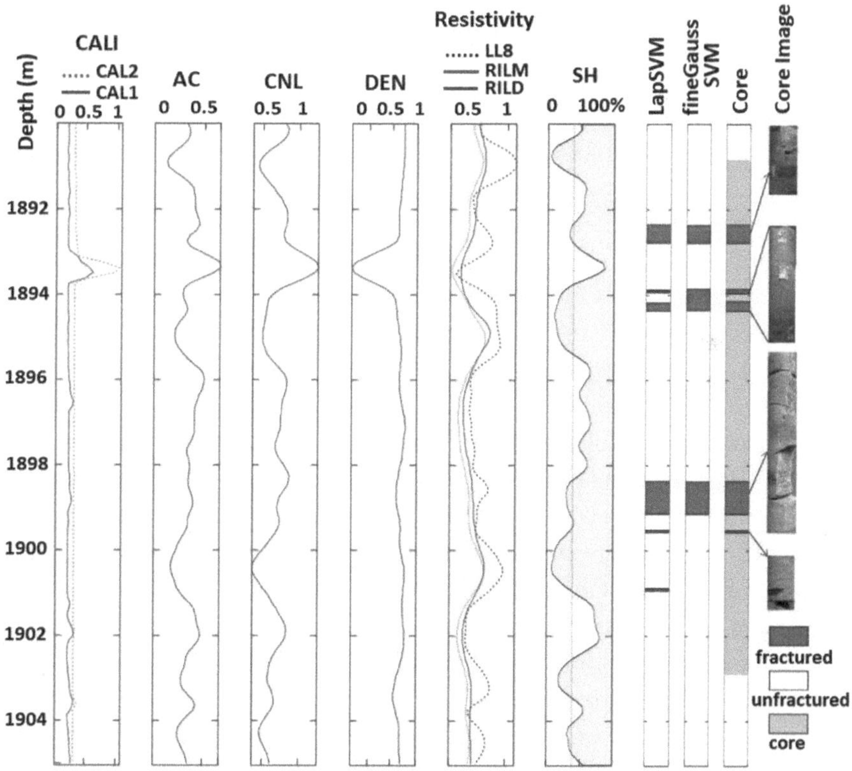

Figure 8–7 Identification of fracture based on conventional well log data by using Laplacian support vector machine and fine Gaussian support vector machine algorithms. Six fractured zones are identified by using Laplacian support vector machine which is in good agreement with core-seen fractures, while fine Gaussian support vector machine predicts only four fractured zones (Dong et al., 2020b).

8.4 Artificial intelligence in petrophysical properties estimation

Multi-stage production from multi-layered tight gas sandstones requires a comprehensive knowledge of petrophysical rock properties. Porosity, permeability, water saturation, and capillary pressure are the main petrophysical properties used to make a geological model of tight gas sandstones. Due to the technical issues such as logging tools failure, wellbore instability issues (e.g., washout and breakouts), and economic considerations, well logging suites are either incomplete or erroneous. Accordingly, AI techniques can be utilized as instrumental tools for the estimation of missing logs.

Porosity: Porosity is an important parameter for reservoir characterization of tight gas sandstones. Unlike permeability, porosity estimation in tight reservoirs can easily be achieved through the interpretation of well logging data such as neutron, sonic, and density logs. Apart from this, an intelligent formulation can be made between the mentioned well logs, also called porosity logs, and core derived porosity. This way, a more accurate porosity estimation will be achieved for a tight gas reservoir.

Permeability: Permeability is a critical parameter for a tight reservoir and its accurate estimation is a major concern to have appropriate planning of well stimulation and development scenarios across a tight gas field. As with the conventional reservoirs, well log data such as gamma-ray, sonic travel time, neutron porosity, density log, and resistivity logs can be considered as inputs to predict permeability by using AI techniques. The low penetration of mud filtrate along with washout issues may cause uncertainties in logging data readings specially resistivity and density logs such that care must be taken in the correctness of input data used as inputs to estimate the reservoir permeability (Baziar et al., 2014). Zhou et al. (2021) stated that the ratio of true formation resistivity to flushed zone resistivity (TXO) can better represent the permeability of a tight reservoir based on the mud filtrate invasion depth (depth of penetration). In the ultra-low permeability tight sandstone, due to the very small penetration depth, the reading of the transition zone is close to the uninvaded zone causing a similar intermediate and deep resistivity reading. In the case of higher permeability, the invasion profile will be deeper causing very close short and intermediate resistivity logs. A sample of estimated permeability in the Middle Jurassic Shaximiao tight gas Formation of the Sichuan Basin is shown in Fig. 8–8.

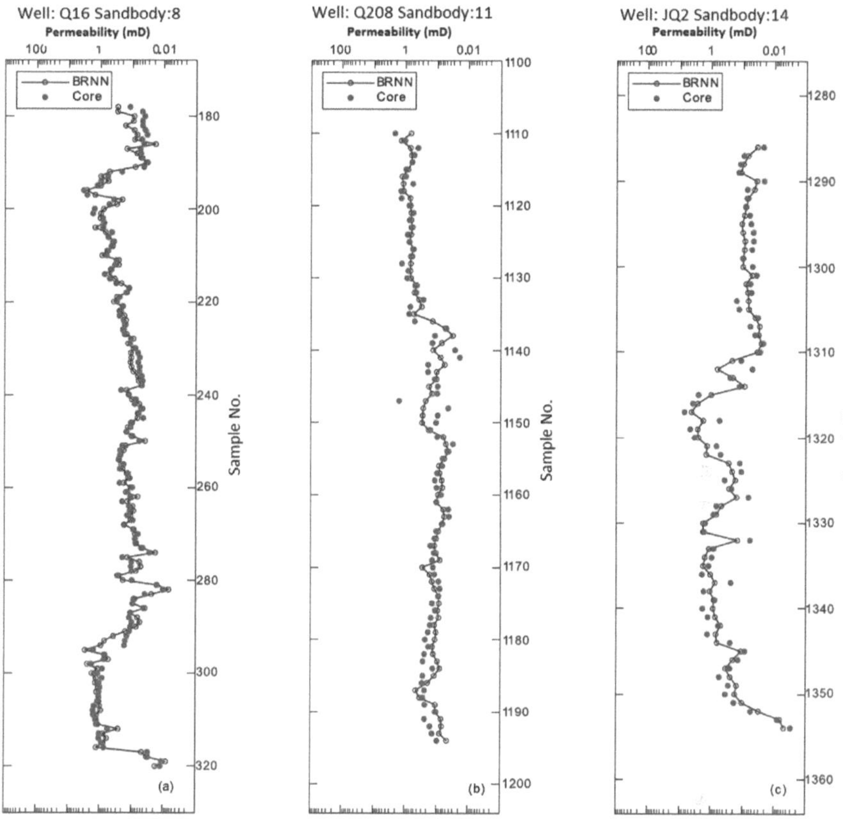

Figure 8–8 A sample of estimated permeability using Bayesian regularization neural network in the Middle Jurassic Shaximiao tight gas formation of Sichuan Basin, wells Q16 (a), Q208 (b) and JQ2 (c) (Zhou et al., 2021).

Water saturation: Accurate estimation of water saturation is a critical issue in the volumetric calculation of tight gas reservoirs. The nanopores present in the structure of tight sandstones make complexities in resistivity logs readings and the corresponding water saturation. Nanopores and microporosities filled with water cause a decrease in deep resistivity reading and hence an increase in water saturation leading to the pessimistic reserve evaluations. Once calibrated with core-derived water saturation, intelligent systems aid in Sw estimation from well logs. An example of water saturation estimation from well log data by using NNs is shown in Fig. 8–9. As is seen, the Imperialist Competitive Algorithm (ICA) optimized NN outperforms Indonesian, improved Indonesian, back-propagation and ICA optimized ANN in estimating core-derived water saturation from conventional well logs (Amiri et al., 2015).

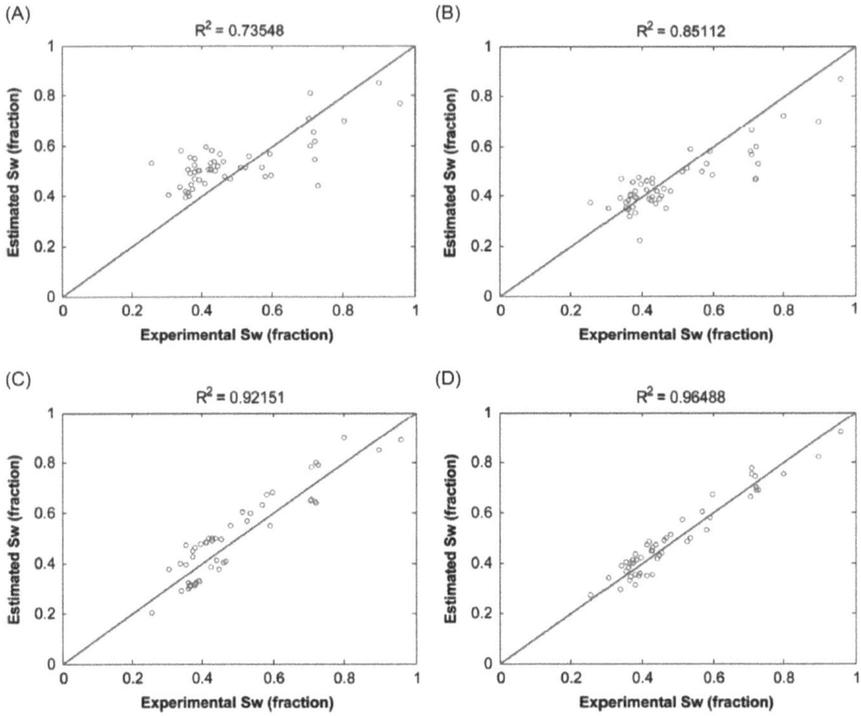

Figure 8–9 Cross-plots comparing core-derived water saturation estimation based on Indonesia (A), improved Indonesia (B), back-propagation neural network (C) and imperialist competitive algorithm optimized ANN (D) from conventional well logs. Imperialist competitive algorithm optimized neural network outperforms other water saturation estimation models (Amiri, et al., 2015).

8.5 Artificial intelligence in seismic reservoir characterization

Using intelligent techniques in seismic data interpretation of tight gas sandstones makes it possible to estimate the spatial distribution of rock properties such as porosity, pore types, water saturation, and facies. In this regard, the multi-attribute transform of seismic data is the basis for the estimation of many rock properties. A seismic attribute is defined as any quantity or property extracted from seismic data. To date, many seismic attributes have been derived from seismic data, but most of them are only mathematical or statistical parameters that have no relation to rock properties. Therefore, a limited number of seismic attributes can be employed for reservoir characterization of tight gas sandstones from which acoustic impedance, instantaneous frequency, instantaneous phase, coherency, and polarity can be mentioned. The main categories of seismic attributes and their application are as follows.

- Frequency attributes: hydrocarbons identification, fluid contacts, fractures zones
- Polarity attributes: lithology identification
- Coherency: identification of faults and fractures
- Phase attributes: identification of reflectors continuity, fractures zones
- Amplitude envelope: recognition of bright spots, tuning effect, bed boundaries, sequence boundaries,
- Spectral decomposition: identification of thin beds.

Acoustic impedance is inversely correlated to the formation of porosity and gas-bearing zones. The low impedance zone is marked by green color in the upper parts of Fig. 8–10 indicates a gas-saturated zone in the tight gas sandstones of the Whicher Range gas field, Perth Basin, Western Australia. An example of porosity derived from NNs by using seismic multi-attribute data is illustrated in Fig. 8–11.

Apart from porosity, pore type identification in a tight gas sandstone is an effective parameter showing suitable areas for gas accumulation and hydraulic fracturing operation. The connected and interconnected pores once combined with the hydraulically created fractures play an important role in gas flow into the production well. Additionally, high porosity and low permeability zones with isolated porosities in a tight gas reservoir are candidate intervals for hydraulic fracturing. The main source of pore types determination is petrographic studies of cores and cuttings. In the absence of core data, petrophysical well logs can be used as an indirect method for

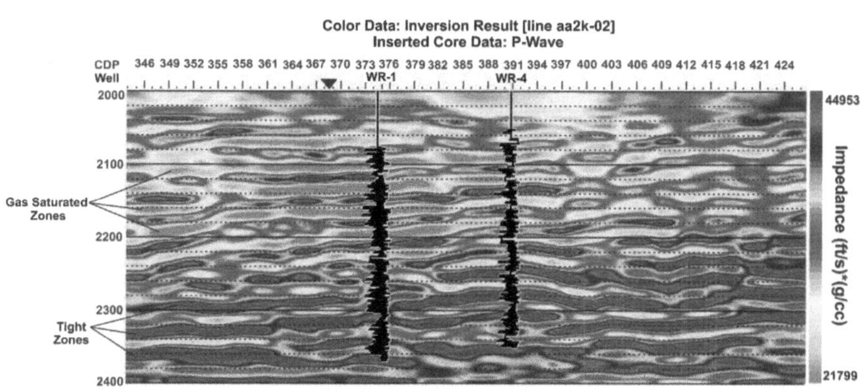

Figure 8–10 Acoustic impedance section from tight gas sandstones of the Whicher Range gas field, Perth Basin, Western Australia (Kadkhodaie & Kadkhodaie, 2022).

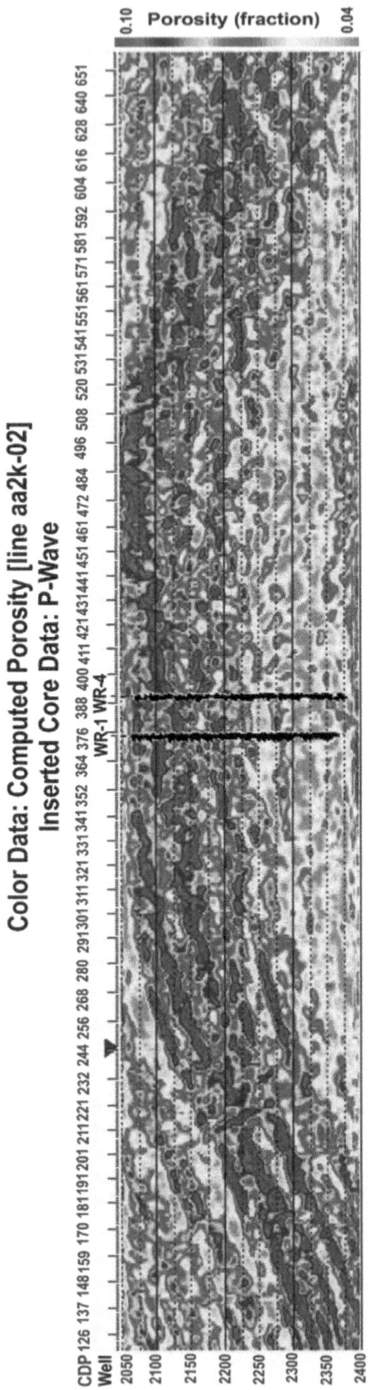

Figure 8–11 Porosity section derived from intelligent transform of seismic attributes in tight gas sandstones of the Whicher Range gas field, Perth Basin, Western Australia (Kadkhodaie & Kadkhodaie, 2022).

pore types identification. The concept of velocity deviation log (VDL) proposed by Anselmetti and Eberli (1999) is an effective approach for pore typing from well-logging data. The VDL as a pore type indicator log is derived by using the following equation.

$$\mathrm{VDL} = (Vp_{\mathrm{real}} - Vp_{\mathrm{synthetic}}) \times 1000 \tag{8.6}$$

$$Vp_{\mathrm{real}} = \frac{304.8}{DT} \tag{8.7}$$

$$Vp_{\mathrm{synthetic}} = \frac{304.8}{DT_{\mathrm{synthetic}}} \tag{8.8}$$

where Vp_{real} (km/s) is sonic velocity derived from Eq. (8.7) and $Vp_{\mathrm{synthetic}}$ is the equivalent sonic velocity (km/s) derived from replacing the total porosity in the Wyllie equation Eq. (8.9) (Wyllie et al., 1956) as follows.

$$\varphi_{\mathrm{total}} = \frac{X - DT_{\mathrm{ma}}}{DT_{fl} - DT_{\mathrm{ma}}} \tag{8.9}$$

where X is the equivalent sonic transit time, also called $DT_{\mathrm{synthetic}}$, for a known porosity. $DT_{\mathrm{synthetic}}$ can easily be converted to velocity by using Eq. (8.8). DT_f and DT_{ma} are the sonic transit time in fluid and matrix, respectively. The total porosity is derived from the neutron, density, or neutron-density porosity. The Wyllie equation was originally developed for porosity calculation from the sonic log. However, the VDL concept quantifies the difference between the velocity derived from sonic transit time and that derived by replacing the total porosity in Wyllie's equation (Wyllie et al., 1956). The sonic log measures the effective porosities, while the neutron and density log measure the total porosity. That is, the velocity corresponding to the difference between the effective and total porosity is quantified as VDL. VDL can be a negative or positive value. Positive deviation (greater than +500 m/s) indicates isolated pore types such as moldic or vuggy pores. Negative deviations (less than −500 m/s) show connected pores and fractures. Zero deviation in the range of −500 to +500 m/s indicates intercrystalline, intergranular, and microporosities.

Although the VDL approach was originally developed for carbonate reservoirs, it has successfully been applied in the case of tight gas sandstone reservoirs. For example, Kadkhodaie & Kadkhodaie (2022) calibrated the VDL to seismic attributes by using NNs for the tight gas sandstones of the

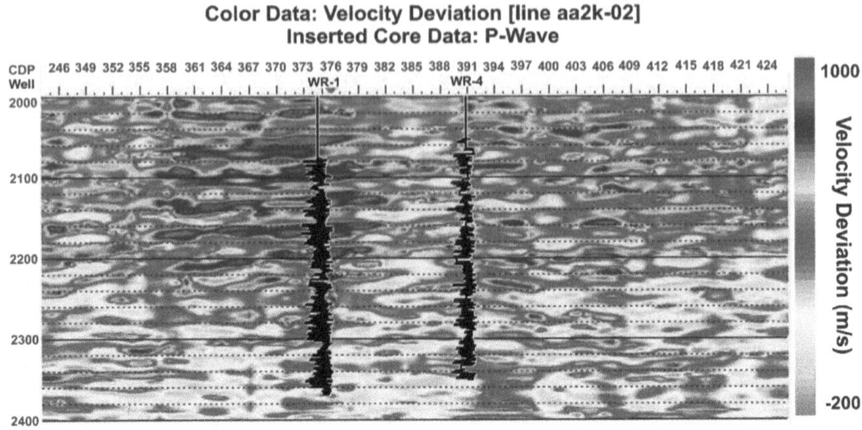

Figure 8–12 A section of pore types derived from the neural network and multi-attribute list in the Whicher Range gas field, Perth Basin, Western Australia. Yellow and green colors (zero deviations) indicate microporosities, while the upper blue colors represent positive deviations with isolated porosities (Kadkhodaie and Kadkhodaie, 2022). (For interpretation of the references to color in this figure legend, the reader is referred to the web version of this chapter).

Willespie Formation (Permian), and in this way they derived 2D sections of reservoir pore types (Fig. 8–12). The results aid in the more successful implementation of hydraulic fracturing through mapping high reservoir quality zones.

References

Amiri, M., Ghiasi-Freez, J., Golkar, B., Hatampour, A., 2015. Improving water saturation estimation in a tight shaly sandstone reservoir using artificial neural network optimized by imperialist competitive algorithm-a case study. Journal of Petroleum Science and Engineering 127, 347–358.

Anselmetti, F., Eberli, G.P., 1999. The velocity deviation log; a tool to predict pore type and permeability trends in carbonate drill holes from sonic and porosity or density logs. American Association of Petroleum Geologists (AAPG) Bulletin 83 (3), 450–466.

Baziar, S., Tadayoni, M., Nabi-Bidhendi, M., Khalili, M., 2014. Prediction of permeability in a tight gas reservoir by using three soft computing approaches: a comparative study. Journal of Natural Gas Science and Engineering 21, 718–724.

Bhatt, A., Helle, H.B., 1999. Porosity, permeability and TOC prediction from well logs using a neural network approach. In: 61st EAGE, Conf., Helsinki, Finland, pp. 1–4.

Bishop, C.M., 1995. Neural Networks for Pattern Recognition. Clarendon Press, Oxford, p. 670.

Bullinaria, J.A., 2008. Neural computation: lecture 19. Course presentation. School of Computer science. University of Birmingham.

Chen, C.H., Lin, Z.S., 2006. A committee machine with empirical formulas for permeability prediction. Computers & Geosciences 32, 485−496.

Chen, M.S., Wang, S.W., 1999. Fuzzy clustering analysis for optimizing fuzzy membership functions. Fuzzy Sets and Systems 103, 239−254.

Chiu, S., 1994. Fuzzy model identification based on cluster estimation. Journal of Intelligent and Fuzzy Systems 2, 267−278.

Dong, S., Zheng, L., Lyu, W., Xia, D., Liu, J., Wu, Y., Du, X., 2020a. Fracture identification and evaluation using conventional logs in tight sandstones: A case study in the Ordos Basin, China. Energy Geoscience 1(3-4), 115−123.

Dong, S., Zheng, L., Lyu, W., Xu, C., Liu, J., Mao, Z., Tian, H., Sun, F., 2020b. Fracture identification by semi-supervised learning using conventional logs in tight sandstones of Ordos Basin, China. Journal of Natural Gas Science and Engineering 76, 103131.

Folk, R.L., 1959. Practical petrographic classification of limestones. American Association of Petroleum Geologists Bulletin 43, 1−38.

Folk, R.L., 1962. Spectral subdivision of limestone types. In: Ham, W.E. (Ed.), Classification of Carbonate Rocks-A Symposium, 1. American Association of Petroleum Geologists Memoir, pp. 62−84.

Golsanami, N., Kadkhodaie, A., Erfani, A., 2015. Synthesis of capillary pressure curves from post stack seismic data with the use of intelligent estimators: a case study from the Iranian part of the South Pars gas field, Persian Gulf Basin. Applied Geophysics 112, 215−225.

Golsanami, N., Kadkhodaie, A., Sharghi, Y., 2014. Estimating NMR T2 distribution data from well log data with the use of a committee machine approach: a case study from the asmari formation in the Zagros Basin, Iran. Journal of Petroleum Science and Engineering 114, 38−51.

Gong, L., Zeng, L., Gao, Z., Zhu, R., Zhang, B., 2016. Reservoir characterization and origin of tight gas sandstones in the Upper Triassic Xujiahe formation, Western Sichuan Basin, China. Journal of Petroleum Exploration and Production Technology 6, 319-329.

Haykin, S., 1999. Neural Networks: A Comprehensive Foundation. Prentice Hall, Englewood Cliffs, NJ, p. 842.

Hinton, G., Deng, L., Yu, D., Dahi, G.E., Mohamed, A.-R., Jaitly, N., et al., 2012. Deep neural networks for acoustic modeling in speech recognition. IEEE Signal Processing Magazine 29, 82−97.

Ja'Fari, A., Kadkhodaie-Ilkhchi, A., Sharghi, Y., Ghanavati, K., 2011. Fracture density estimation from petrophysical log data using the adaptive neuro-fuzzy inference system. Journal of Geophysics and Engineering 9 (1), 105−114.

Jafari, A., Kadkhodaie, A., Sharghi, Y., Ghaedi, M., 2014. Integration of adaptive neuro-fuzzy inference system, neural networks and geostatistical methods for fracture density modeling. Oil & Gas Science and Technology − Rev. IFP Energies nouvelles 69 (7), 1143−1154.

Ji, J., Lin, M., Jiang, W., Cao, G., Xu, Z., Hao, F., 2022. An improved rock typing method for tight sandstone based on new rock typing indexes and the weighted fuzzy kNN algorithm. Journal of Petroleum Science and Engineering 109956.

Kadkhodaie, A., Kadkhodaie, R., 2022. Acoustic, density, and seismic attribute analysis to aid gas detection and delineation of reservoir properties. In: Wood, D.A., Cai, J. (Eds.), Sustainable Geoscience for Natural Gas Sub-Surface Systems. Elsevier, pp. 51−92. Chapter 3.

Kadkhodaie, A., Rahimpour-Bonab, H., Rezaee, M.R., 2009a. A committee machine with intelligent systems for estimation of total organic carbon content from petrophysical data: an example from the Kangan and Dalan Reservoirs in South Pars Gas Field, Iran. Computers & Geosciences 35, 459–474.

Kadkhodaie, A., Rezaee, M.R., Rahimpour-Bonab, H., 2009b. A committee neural network for prediction of normalized oil content from well log data: an example from South Pars Gas Field, Persian Gulf. Journal of Petroleum Science and Engineering 65, 23–39.

Kamali, M.R., Mirshady, A.A., 2004. Total organic carbon content determined from well logs using ΔLogR and neuro fuzzy techniques. Journal of Petroleum Science and Engineering 45, 141–148.

Krizhevsky, A., Sutskever, I., Hinton, G., 2012. ImageNet classification with deep convolutional neural networks. Proceedings Advances in Neural Information Processing Systems 25, 1090–1098.

Liu, J.-J., Liu, J.-C., 2022. Integrating deep learning and logging data analytics for lithofacies classification and 3D modeling of tight sandstone reservoirs. Geoscience Frontiers 13 (1), 101311.

Lu, Y, Liu, K., Wang, Y., 2020. Identification of Diagenetic Facies in Low-Permeability Sandstone Reservoirs Based on Self-Organizing-Map Neural Network Algorithm. Paper presented at the International Petroleum Technology Conference, Dhahran, Kingdom of Saudi Arabia, January 2020.

Lyu, W.Y., Zeng, L.B., Liu, Z.Q., Liu, G.P., Zu, K.W., 2016. Fracture responses of conventional logs in tight-oil sandstones: a case study of the upper triassic yanchang formation in southwest Ordos Basin, China. AAPG (American Association of Petroleum Geologists) Bulletin 100 (9), 1399–1417.

Lyu, W.Y., Zeng, L.B., Zhou, S.B., Du, X.S., Xia, D.L., Liu, G.P., et al., 2019. Natural fractures in tight-oil sandstones: a case study of the upper triassic yanchang formation in the southwestern Ordos Basin, China. AAPG (American Association of Petroleum Geologists) Bulletin 103 (10), 2343–2367.

Nikravesh, M., Aminzadeh, F., 2003. Soft computing and intelligent data analysis in oil exploration, part 1: Introduction. Fundamentals of Soft Computing. Elsevier, Berkeley, CA, p. 744.

Nouri-Taleghani, M., Kadkhodaie, A., Karimi-Khaledi, M., 2015. Determining hydraulic flow units using a hybrid neural network and multi-resolution graph-based clustering method: case study from South Pars Gasfied, Iran. Journal of Petroleum Geology 38 (2), 177–191.

Pettijohn, F.J., Potter, P.E., Siever, R., 1987. Sand and Sandstones, second ed. Springer Verlag, Germany, p. 553.

Rezaee, M.R., Kadkhodaie, A., Barabadi, A., 2007. Prediction of shear wave velocity from petrophysical data using intelligent systems, a sandstone reservoir of Carnarvon Basin. Journal of Petroleum Science and Engineering 55, 201–212.

Rezaee, M.R., Kadkhodaie-Ilkhchi, A., Mohammad Alizadeh, P., 2008. Intelligent approaches for the synthesis of petrophysical logs. Journal of Geophysics and Engineering 5 (1), 12–26.

Safaei-Farouji, M., Kadkhodaie, A., 2022. Application of ensemble machine learning methods for kerogen type estimation from petrophysical well logs. Journal of Petroleum Science and Engineering 208, 109455.

Sharkey, A.J.C., 1996. On combining artificial neural nets. Connection Science 8, 299–314.

Tokhmchi, B., Memarian, H., Rezaee, M.R., 2010. Estimation of the fracture density in fractured zones using petrophysical logs. Journal of Petroleum Science and Engineering 72, 206–213.

Wang, L., Feng, H.-Y., Cai, N., 2003. Architecture design for distributed process planning. Journal of Manufacturing Systems 22 (2), 99–115.
Wyllie, M.R.J., Gregory, A.R., Gardner, L.W., 1956. Elastic wave velocities in heterogeneous and porous media. Geophysics 21, 41–70.
Zahmatkesh, I., Soleimani, B., Kadkhodaie, A., Golalzadeh, A., Abdollahi, A.-M., 2017. Estimation of DSI log parameters from conventional well log data using a hybrid particle swarm optimization–adaptive neuro-fuzzy inference system. Journal of Petroleum Science and Engineering 157, 842–859.
Zeidenberg, M., 1990. Neural Network in Artificial Intelligence Ellis Horwood Limited, NewYork.
Zhou, Y., Zhao, X., Jiang, C., Liu, S., Han, Z., Wang, G., 2021. Permeability prediction of multi-stage tight gas sandstones based on bayesian regularization neural network. Marine and Petroleum Geology 133, 105320.

CHAPTER 9

Tight sands reservoir modeling

9.1 Introduction

Predicting and knowing the spatial distribution of lithological and petrophysical characteristics of reservoirs is essential for reserve estimation of a tight gas field and more successful implementation of development and production plans (Al-Khalifah and Makkawi, 2002; Izadi and Ghalambor, 2012). Nowadays, spatial distribution and estimation of reservoir characteristics are usually done by geostatistical methods. In fact, geostatistics is a tool considered by geologists and engineers to analyze data and use the results for reservoir forecasting and modeling purposes (Al-Khalifah and Makkawi, 2002). Modeling is the creation of a three-dimensional digital image of a hydrocarbon reservoir in-depth, which includes the reservoir framework (faults and stratigraphic surfaces), the internal structure (sedimentary facies), and petrophysical properties (porosity and permeability) and fluid distribution (water saturation).

Investigation of reservoir properties of tight sandstones in a geocellular model provides a comprehensive understanding of the spatial distribution of reservoir zones and rock types. Also, the integration of reservoir properties (e.g., porosity, permeability, water saturation, etc.) and depositional and diagenetic characteristics of reservoir rocks in the model framework, helps to evaluate the main factors controlling the reservoir heterogeneity and production zones in the field. The accuracy and reliability of the model results depend basically on the data and algorithms used in the modeling process. Therefore, quantitative and qualitative control of geological, petrophysical, and seismic data, is an important and initial step that should be considered in reservoir modeling. In a certain field or basin, different data sources, including core, well log, seismic, and reservoir engineering parameters of them cover specific parts of a model, maybe in incomplete, missing, or unreliable forms. Missing and incomplete data are related to out-of-reach or unavailable data in some wells or parts of the study field or area. For example, core data or some specific well logs such as FMI logs are not available in all wells of a field. Moreover, in some areas, due to the shallow drilling depth or geological conditions such as

Reservoir Characterization of Tight Gas Sandstones
DOI: https://doi.org/10.1016/B978-0-323-90180-2.00009-8

the effect of erosion, all or a part of the studied interval may be missed. Also, unreliable data that are primarily derived from the impact of geological and environmental conditions on measuring instruments underground include all data with low quality and invalid values. Log readings in intervals with breakout and washout, and also seismic data with low quality in deep intervals, are examples of unreliable data. Furthermore, some data such as interpreted well logs (e.g., effective porosity, the volume of shale, water saturation) and rock types (e.g., electrofacies and flow units) are interpreted and processed data that their quality and reliability depend on the methods and parameters used in data analysis. Therefore, in a field with many drilled wells, it is essential to select key wells based on the quantity and quality of available data. It should also be noted that the selection of key wells varies according to the study targets and the number of drilled wells, and in addition to data, other parameters such as well spacing and drilling depth should also be considered. On the other hand, the choice of a suitable algorithm as well as the statistical parameters of the model such as the variogram ranges, significantly depends on the reservoir heterogeneity and the data and information used, whose resolution (quantity and quality) reveals this heterogeneity. Certainly, a comprehensive understanding of reservoir characteristics requires integrating the results of all data sources, including geology, petrophysics, and reservoir engineering.

Core description and core analysis tests provide the most important data and information on reservoir rocks. For this purpose, macroscopic description of core intervals (core facies) and microscopic observations of thin sections (petrofacies) are used for the description of lithology, depositional facies, and diagenetic features of tight sandstone reservoirs. The results of these studies in the context of sedimentary rock types or diagenetic facies, provide basic data and information for the description and interpretation of petrophysical rock types. In most tight sandstone reservoirs, it is difficult to correlate microscopic facies (petrofacies) and even diagenetic facies with good logs and to map them in the model framework. On the other hand, extracting a complex facies microscale model will not be useful for use in dynamic reservoir models. Also, in most cases, due to the effect of diagenesis on the pore system of reservoir rocks, different petrofacies, cannot be differentiated based on the primary reservoir quality, and overlap between them is observed on the core porosity and permeability plot. Therefore, due to the strong heterogeneity and complexity in microscale reservoir characteristics, the description and differentiation of facies based on the large-scale textural characteristics and lithology (lithofacis), which can also be correlated with well logs, is more practical for reservoir modeling purposes than petrofacis. Tight sandstones of the Zeivar Formation of Moghan Basin in northwestern Iran and also tight sandstone of the Willespie Formation in the Perth Basin of Western Australia are examples of tight reservoirs with heterolithic lithology such that their depositional texture changes from very fine and silty to very coarse and granular in size. Also, variations in the spatial distribution of sedimentary units in terms of lithology and thickness is a characteristic of clastic rocks that is related to the tectonic-sedimentary evolution of the basin and can be revealed by isopach and isolith maps. Therefore, the application of these maps as well as a stratigraphic correlation between wells in a tight gas field will help to achieve an understanding of the lateral distribution of lithofacies and also identify the depocenters (areas of maximum deposition) of the basin. The results of these studies provide an overview of the sedimentary basin framework, based on which the modeling process can intelligently be managed. Fig. 9–1 demonstrates an isopach map of the Zeivar tight sandstone Formation of Moghan Basin in northwestern Iran, which is based on the thickness data obtained from the surface sections (outcrops) of the area. As is seen, the depocenters related to this formation are concentrated in the northern parts of the basin. Fig. 9–2 also shows an isolith map of conglomerate for the lower zone of this formation, which indicates an overall decrease in conglomerate lithofacies from south to north of the basin. Moreover, a stratigraphic correlation between the surface and subsurface

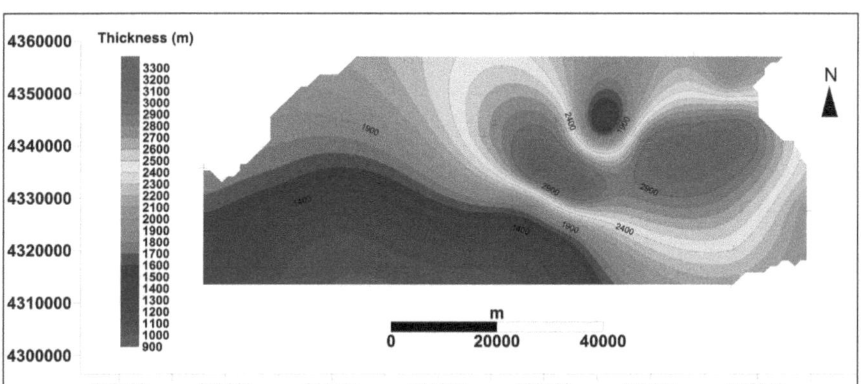

Figure 9–1 Isopach map of the Zeivar tight sandstone Formation in the Moghan area of northwestern Iran. The location of surface sections (outcrops) is marked on the map with plus signs.

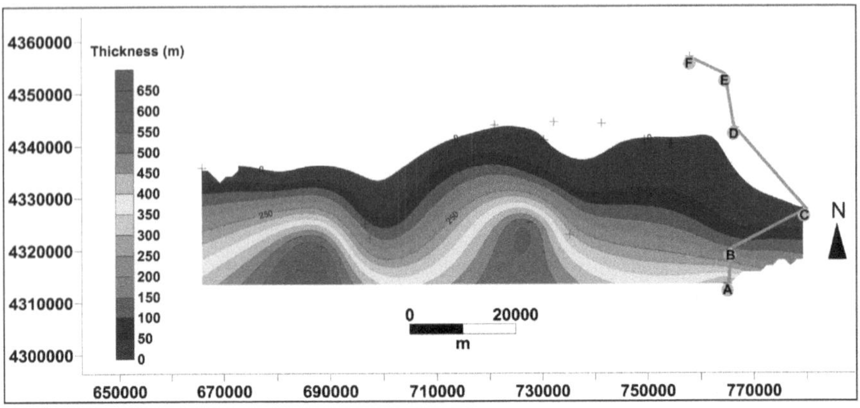

Figure 9–2 Isolith map of conglomerate lithofacies for the lower zone of the Zeivar Formation of Moghan area in northwestern Iran. The location of surface sections (outcrops) is marked on the map with plus signs.

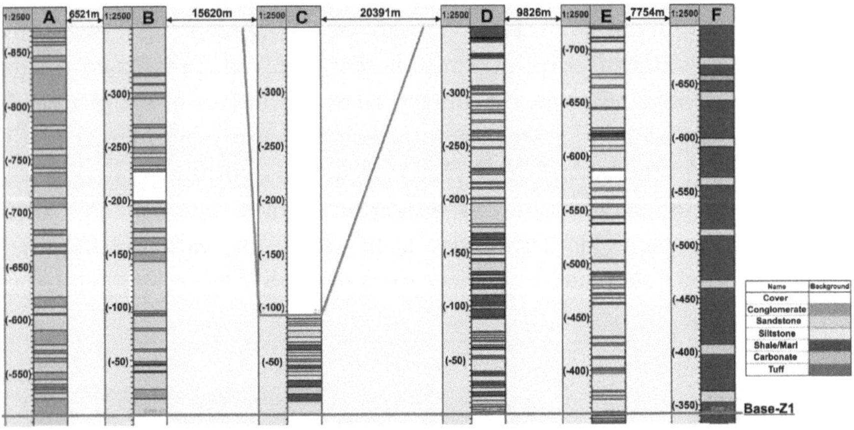

Figure 9–3 Stratigraphic correlation for the lower zone of the Zeivar Formation in one cross-section (AB) of the study area. The cross-section location is shown in Fig. 9–2.

sections for this zone in one cross-section (AB), as shown in Fig. 9–3, indicates the dominance of conglomerate in the southern sections and the absence of these lithofacies in the northern sections of the study area. The decrease and disappearance of coarse-grained lithofacies (conglomerate as well as sandstone) in the northern sections are associated with the predominance of fine-grained facies (siltstone and shale/marl) in these sections. This refers to the sediment provenance in southern parts of the study basin, as mentioned in previous studies (i.e., Lukoil Report, 2007).

Core analysis data, including CCAL (e.g., porosity and permeability) and SCAL (e.g., relative permeability, capillary pressure, and water saturation) are used for pore size distribution analysis and differentiation of flow unit-based rock types. In tight sandstone reservoirs, pore throat size is considered an important parameter for the evaluation of reservoir quality (Wu et al., 2018). Moreover, flow units as an indicator of the pore system potential for fluid flow can be considered in dynamic reservoir modeling. Also, raw and interpreted well logs can be used efficiently to differentiate sandy units, reservoir zones, and log-based rock types (electrofacies).

In most cases, facies modeling based on log-derived facies (electrofacies) is preferred to core facies (petrofacies and lithofacies) because, firstly, the electrofaces are continuous along with the reservoir interval, and secondly, they can be well correlated between the wells of a studied field. The optimal number of electrofacies, which mainly depends on the reservoir heterogeneity, is determined by correlating with the core facies. Gamma-ray log (GR) and porosity well logs including neutron (NPHI), density (RHOB), and sonic (DT) are the most important logs used in the differentiation of electrofacies. However, intervals with washout and breakout problems severely affect NPHI and RHOB log responses, leading to unreliable values log readings. GR is a useful log in the differentiation of sandy and shaly units. Kadkhodaie-Ilkhchi et al. (2013) by cluster analysis of good logs, used GR in sedimentary rock typing of tight sandstone intervals of the Whicher Range field of the Perth Basin in Western Australia. Consequently, they recognized three sedimentary rock types, including clean sandstone (GR < 80 API), dirty sands (80 < GR < 130 API), and shaly units (GR > 130 API).

Seismic data characterize the reservoirs on a large scale, thus they cover data and information gaps between the wells. In addition to the depiction of structural features (faults and folds) and analyzing stratal surfaces, thickness, and continuity of subsurface rock units, seismic attributes are effectively used to interpret and model reservoir characteristics. Although the frequency of seismic data is lower and their scale is larger than cores and well logs, they can be correlated and integrated with good logs by seismic inversion. Acoustic impedance (AI), as a seismic attribute, is the final result of seismic inversion that shows an inverse relationship with porosity, thus it can be used to estimate the porosity from seismic data as a secondary variable.

Workflow for static reservoir modeling consists of two main steps, including structural and property modeling. Structural modeling establishes the reservoir framework related to faults and stratal surfaces and results in a three-dimensional geocellular model with many grid cells of certain sizes. Creating this model is essential for geostatistical methods to estimate reservoir properties cell by cell and show their spatial distribution in the study area. Property modeling including facies models (e.g., lithofacies, and electrofacies) and petrophysical models (e.g., porosity, permeability, and water saturation) illustrates the distribution of reservoir properties within the geocellular skeleton by upscaling and variogram analysis. Upscaling is the propagation of reservoir properties to grid cells by which a certain value of the desired parameter is assigned to each cell. Also, the study of the spatial relationship of data in horizontal and vertical directions is accomplished through variogram analysis. The main direction of the variogram corresponds to the direction with the least variation on the variogram map indicating the variability in sedimentary characteristics of reservoir rocks. Fig. 9–4 shows examples of variogram analysis of permeability data in the major and minor directions.

The algorithms used in the modeling of facies and petrophysical characteristics of the reservoir are deterministic and stochastic. Deterministic methods such as kriging and cokriging create only one model and are not reliable in areas without data. But stochastic methods such as Gaussian simulation and Gaussian co-simulation produce more models and their use in estimating and distributing the desired parameter leads to more realistic results than what is seen in deterministic methods. Facies modeling is

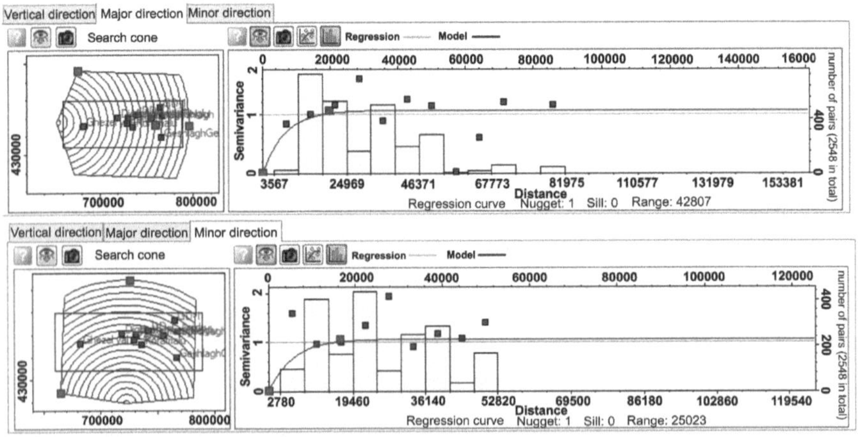

Figure 9–4 Variogram analysis for permeability data in major and minor directions for tight sandstones of the Zeivar tight sandstone Formation in Moghan basin, northwest of Iran.

accomplished by object-based and pixel-based algorithms. The commonly used pixel-based algorithm for facies modeling is Sequential Indicator Simulation (SIS), and for petrophysical models such as porosity and permeability, Sequential Gaussian Simulation (SGS) algorithm is used.

9.2 Facies modeling

Investigation of facies in the framework of the 3D model, in addition to depicting the reservoir heterogeneity related to the sedimentary environment, can be used as a complementary parameter for petrophysical modeling to provide the spatial distribution of reservoir properties with high accuracy. Furthermore, due to the different degrees of reservoir heterogeneity, the sedimentary facies or lithofacies have considerable control over petrophysical properties (Ma et al., 2009). Therefore, the results of a facies-conditioned petrophysical model are more reliable than the results of a model produced based only on variogram analysis of data related to that specific petrophysical parameter. Wei et al. (2016) indicated the importance of sedimentary facies and diagenesis on the porosity evolution of tight gas sandstones in the Liaohe Basin of China, and the results were used for modeling and prediction of porosity in the study area. Also, Orsini and Rezaee (2012) and Kadkhodaie-Ilkhchi et al. (2019) indicated the importance of facies and depositional environment on reservoir properties and controlling the large-scale reservoir heterogeneity of tight gas sandstones of the Perth Basin in Western Australia.

9.3 Petrophysical modeling

Petrophysical parameters such as porosity and permeability, unlike facies that are discrete parameters, are continuous and variations in these parameters can be related to the depositional, diagenetic, and even seismic characteristics of reservoir rocks. Therefore, these characteristics can be considered and involved as a key control in the petrophysical modeling process to decrease the model uncertainty and reach more realistic results. This is especially important in tight sandstone reservoirs with a highly heterogeneous nature and complex pore system. For example, a porosity model could be conditioned to different facies of sedimentary/diagenetic rock types recognized through the description of core intervals or well log analysis. Similarly, due to the inverse relationship between the AI and porosity, this parameter can be used in the modeling process to estimate porosity using the cokriging method. In addition, by establishing the relationship between petrophysical parameters such as porosity, permeability, and fluid saturation in tight sandstones, it is possible to estimate one parameter from another. For example, an accurate model of permeability could be extracted from the correlation between porosity and permeability, which the results have an impact on the hydrocarbon recovery rate. Also, the relationship between porosity and water saturation could be used for the prediction and accurate modeling of water saturation such that the results have an impact on the estimation of the in-place volumetrics (Moore et al., 2016).

9.4 Example of reservoir modeling for tight gas sandstone reservoirs

In this section, as an example of unconventional reservoirs, facies and petrophysical modeling for tight sandstones of the Whicher Range field in the Perth Basin of Western Australia are briefly discussed. The produced models are based on core and well log data from five drilled wells (WR-1 to WR-5) of the field. Fig. 9–5 shows the three-dimensional network structure of the field, which is a faulted anticline with a rough surface. Facies modeling was extracted based on the information obtained from the description of core intervals and differentiation of sedimentary rock types based on GR log responses. As a result, three main rock types, including clean sand (GR < 80 API), fine and silty sand (80 < GR < 130

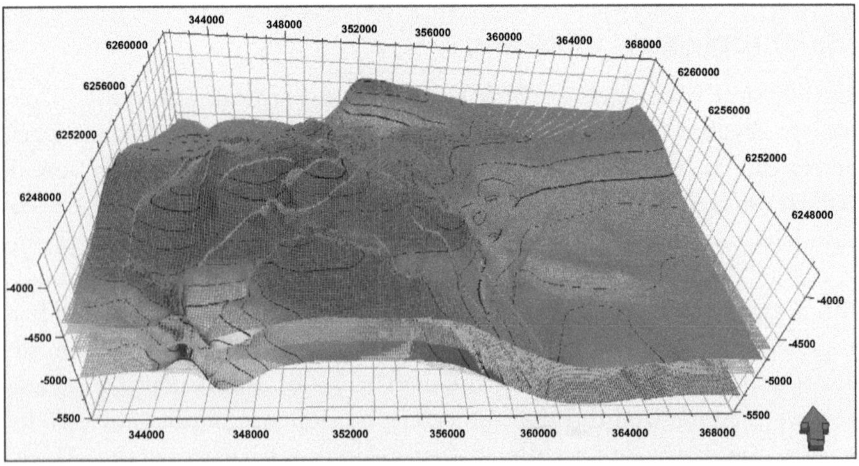

Figure 9–5 Structural framework of the Whicher Range Field in the Perth Basin.

API), and shaly units (GR > 130 API) were identified and used for facies modeling. Fig. 9–6A demonstrates a 3D facies model extracted by SIS algorithm, and Fig. 9–6B shows a cross-section of this model along with the studied wells. The results of the model can be compared with good logs and upscaled cells (Fig. 9–7). The small difference between them indicates a good estimate of the distribution of facies within the

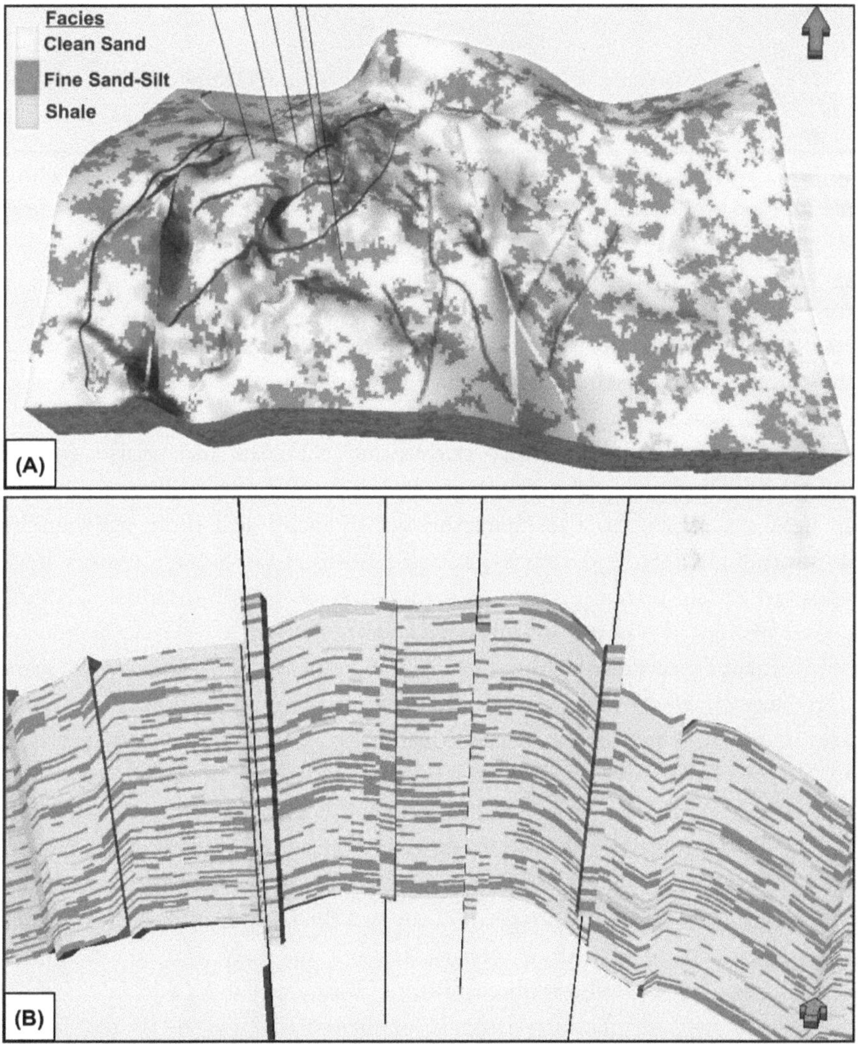

Figure 9–6 (A) 3D facies model of tight sandstones of the Whicher Range field in the Perth Basin. (B) A Cross-section with the E-W trend of the model.

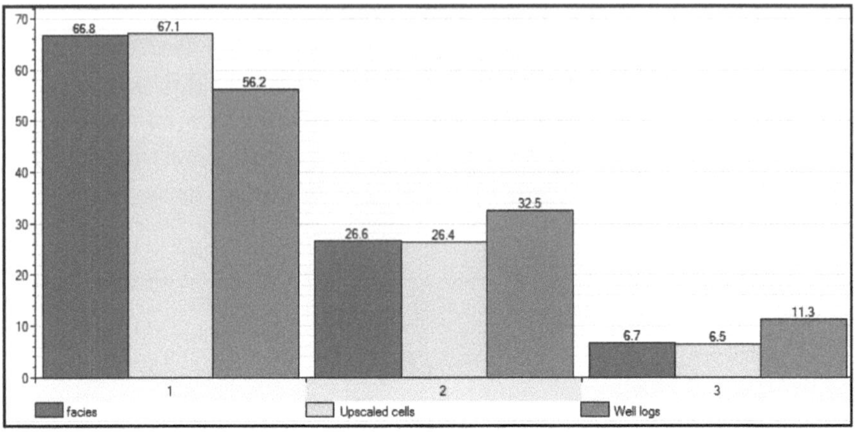

Figure 9–7 Comparison of facies frequency estimated from the model with that from well log and upscaled cells in the tight sandstones of the Whicher Range field, Perth Basin.

three-dimensional network of the model. According to these figures, clean sandstones as the main reservoir units alternate with fine-grained sandstones, siltstones and shale units. Also, the thickness of the reservoir facies, as shown in Fig. 9–8, changes from less than 10 to 50 m, and the high thicknesses and continuity correspond to clean and coarse-grained sandstones. Changes in the thickness and continuity of sandstone units in the field are related to the characteristics of facies and their sedimentary environment. Clean and coarse-grained sandstone facies have mainly been deposited in high-energy parts of a fluvial system, as individual channels or a set of high-thickness accumulated channels.

Investigation of shale volume and its variations in the field provides some information on reservoir heterogeneity related to the distribution of sandstone facies under the effect of the depositional environment. Fig. 9–9A displays a 3D model of shale volume derived from the GR log in each well of the field. A study of shale volume along with the studied wells, as shown in Fig. 9–9B, demonstrates that clean and shaly units have different distributions between the studied wells. As WR-1 and WR-4 wells, and to some extent WR-5 include a high percentage of clean and thick sandstones mostly related to the fluvial channels in their interval. In contrast, shaly units are dominant in WR-2 and WR-3 where thin sandstone units are developed.

Porosity and permeability models, as shown in Figs. 9–10 and 9–12, were generated by SGS algorithm. According to a cross-section of the porosity model (Fig. 9–10B), a general increase in porosity is observed

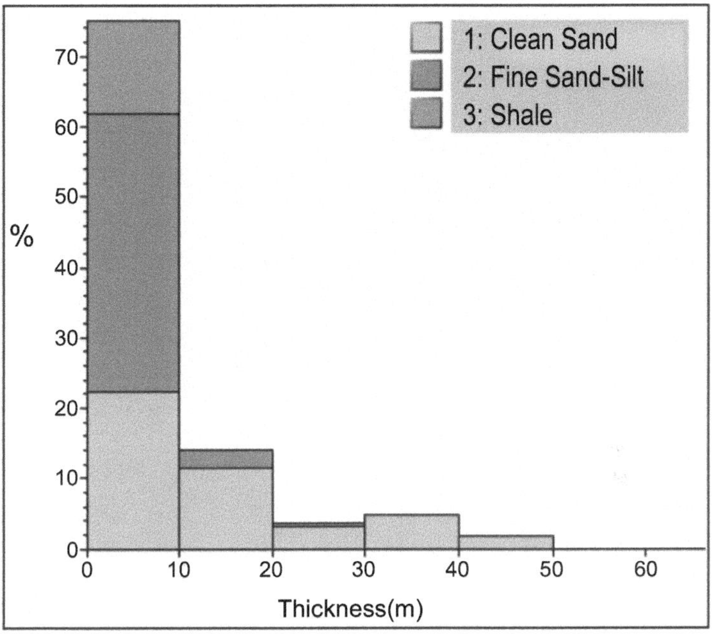

Figure 9–8 Variability in thickness for different lithofacies in tight sandstones of the Whicher Range field, Perth Basin.

towards the upper parts of the formation. This is consistent with the results of seismic inversion studies, according to which the AI shows an inverse trend with respect to porosity, and decreases towards the upper parts of the formation (Kadkhodaie-Ilkhchi et al., 2014). In spite of porosity that shows a wide range, permeability has a narrow range for tight sandstones of the field. This is consistent with the tight nature of sandstone reservoirs and their pore system with the predominance of poorly-connected pores and micropores. Therefore, due to the low values and narrow range of permeability, it is difficult to depict this parameter in a 3D model such that the uncertainty in the model results is high. As most of the porosity in these reservoirs is not effective, it is not a good parameter for estimating permeability. However, permeability shows an inverse relationship with shale volume in the study field (Fig. 9–11). So, the permeability model was extracted based on the correlation between permeability and shale volume (Fig. 9–12A).

According to a cross-section of the model (Fig. 9–12B), although permeability shows slight vertical and lateral changes within the reservoir interval, the presence of some high-permeability zones within the reservoir interval is

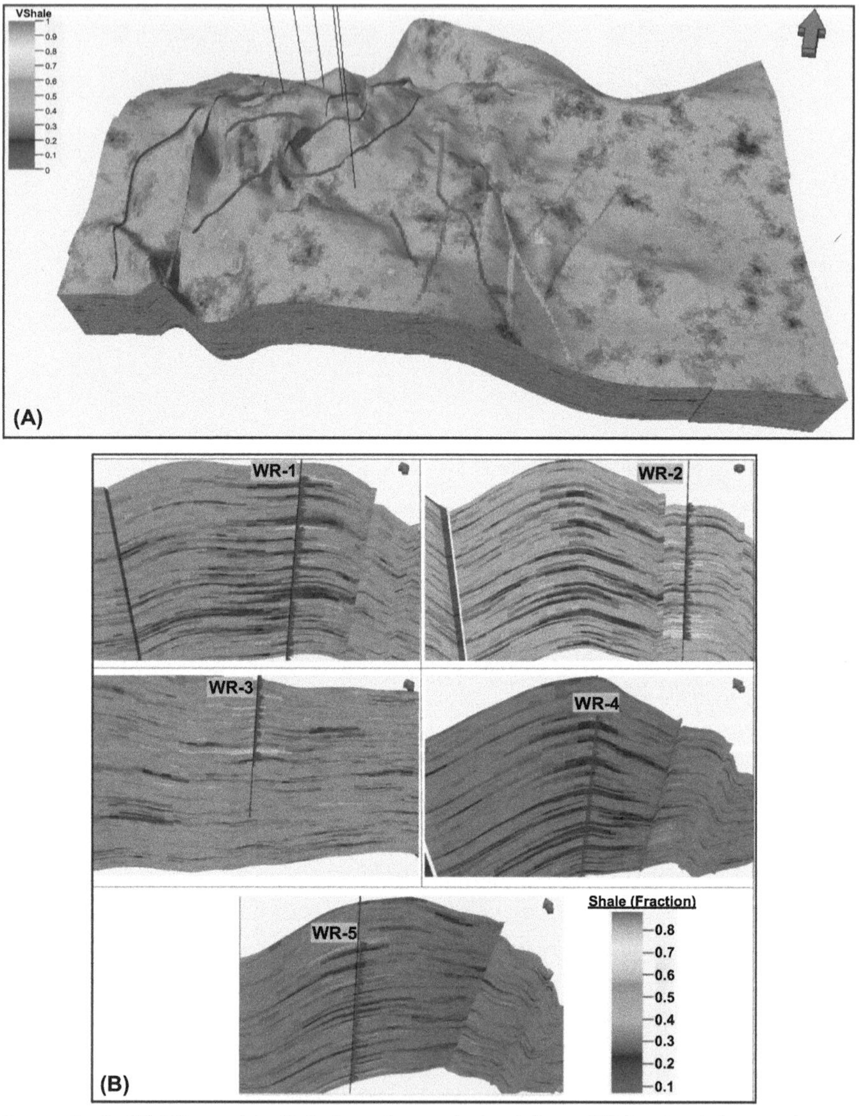

Figure 9–9 (A) 3D model of shale volume derived from GR log in tight sandstones of the Whicher Range field. (B) A cross-section of shale volume model along with the studied wells.

evident. These zones that are considered gas saturated zones in tight sandstones of the field show a close relationship with depositional, diagenetic, and pore system properties of reservoir rocks (Kadkhodaie et al., 2021; Kadkhodaie and Kadkhodaie, 2022).

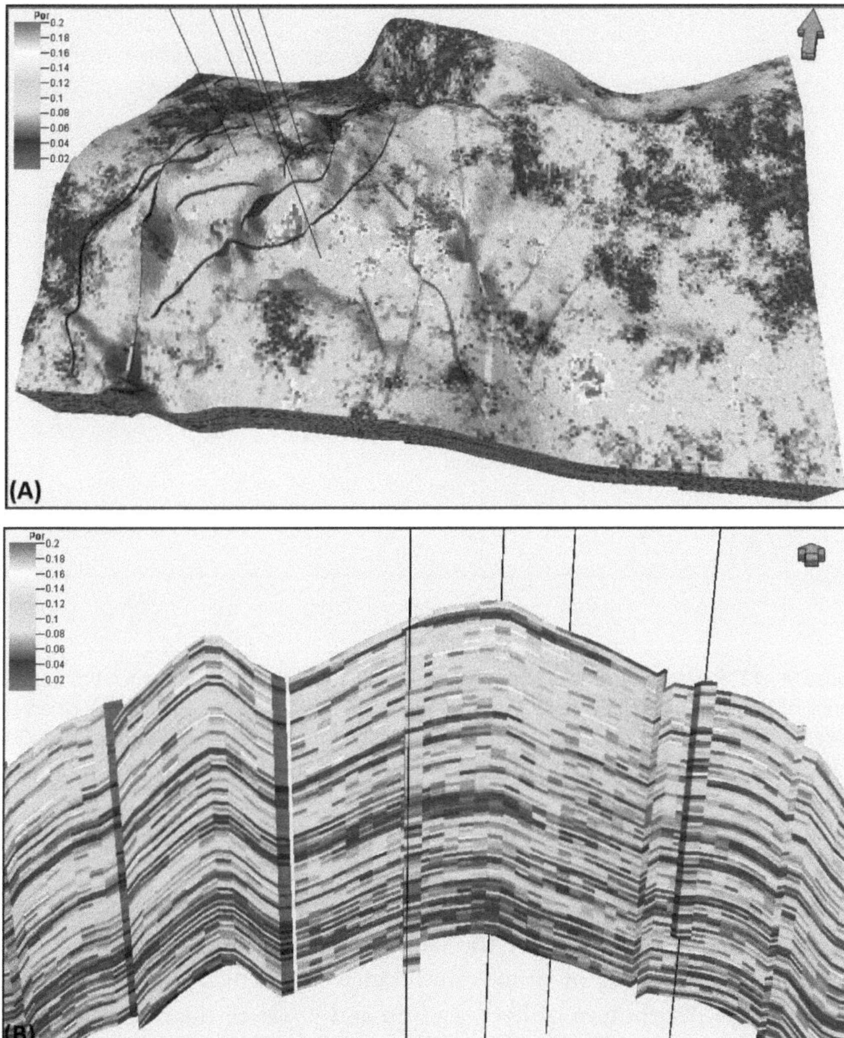

Figure 9–10 (A) 3D model of porosity in tight sandstones of the Whicher Range field. (B) Cross-section of porosity model along with the studied wells.

Making a comparison among porosity, shale volume and permeability models shows that reservoir characteristics of tight sandstones of the Whicher Range field are essentially controlled by depositional facies such that high energy and coarse-grained facies of fluvial channels still form the production zones of the field.

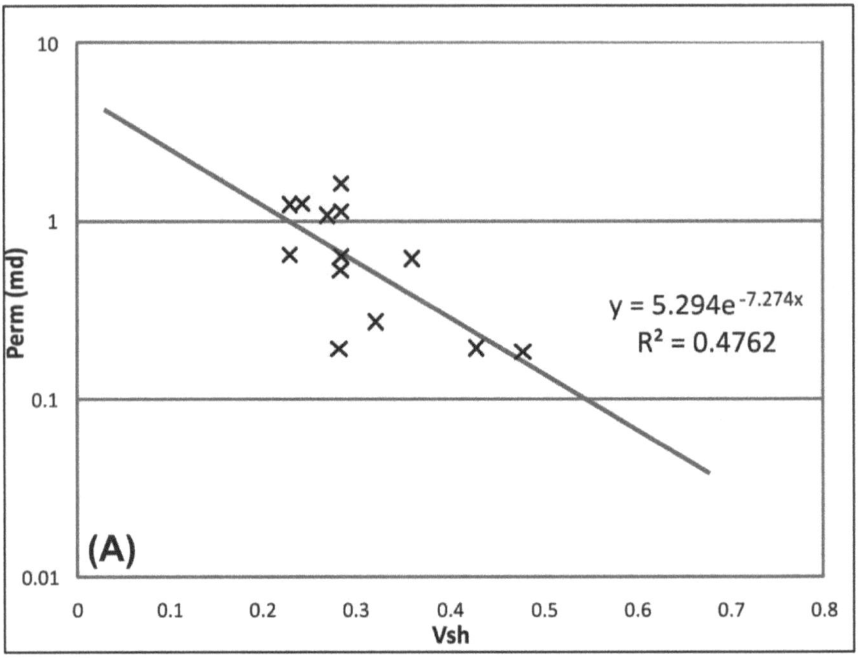

Figure 9–11 Inverse relationship between permeability and shale volume for tight sandstones of the Whicher Range field (Ciftci, 2012). *After Ciftci, B., 2012. First order reservoir modeling of the Whicher Range Field tight gas sands. In: Whicher Range Tight Gas Sands Study, Western Australian Energy Research Alliance, Report 112, pp. 288–307.*

9.5 Height function

Variation in water saturation as a function of the height above the free water level (FWL) is shown by the saturation height function. The saturation height function is of prime importance to calculate the reserve and modeling the distribution of hydrocarbon and water throughout the reservoir interval (Jamiolahmady et al., 2007; Cuddy et al., 2017). The vertical distribution and mobility of the fluids in the reservoir can be investigated by core-derived capillary pressure and relative permeability. The saturation height function derived from capillary pressure curves assigned to each rock type is fundamental data for the construction and conditioning of the water saturation model. Accordingly, defining rock types in a more meaningful and integrated way will lead to reducing uncertainty about the calculation of oil and gas in place.

The low permeability of tight sandstones significantly restricts the gas and water flow, and there is a broad range of saturations and a wide

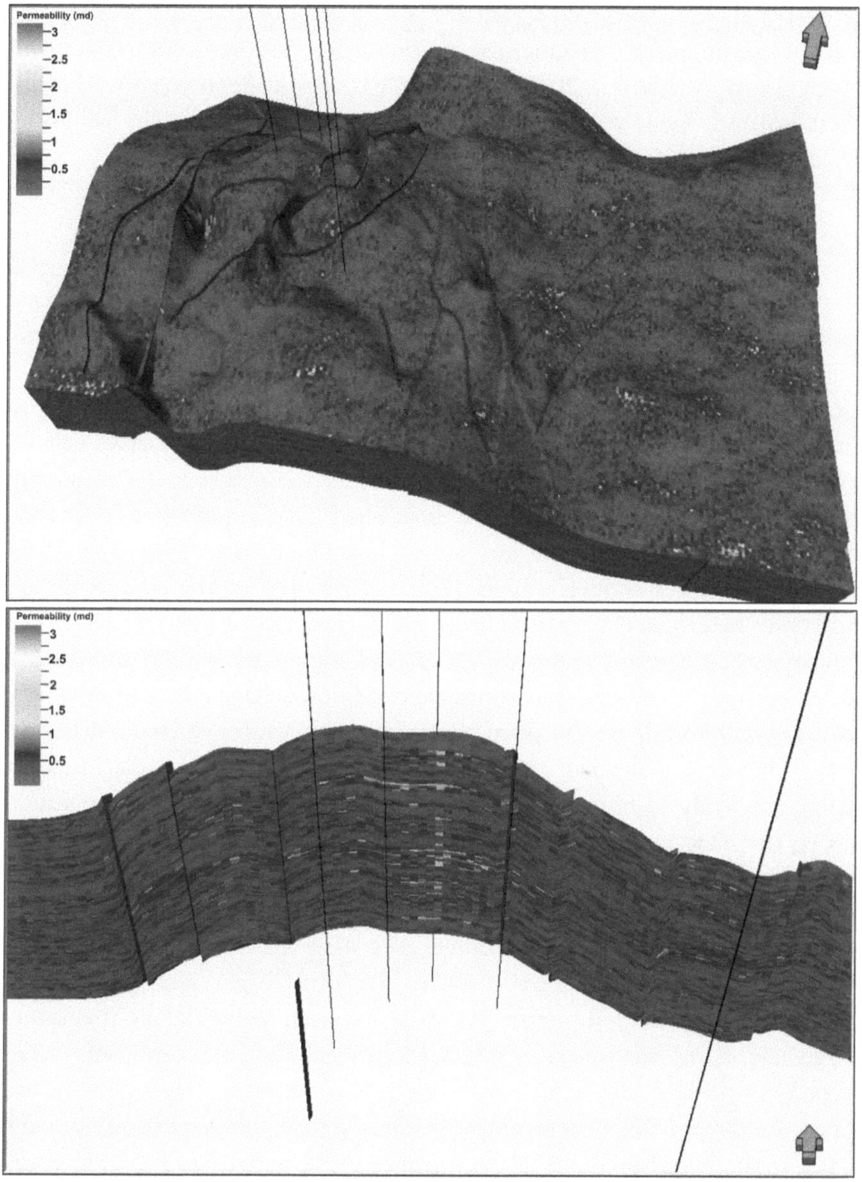

Figure 9–12 (A) 3D model of permeability in tight sandstones of the Whicher Range field. (B) Cross-section of permeability model along with the studied wells.

region above FWL with little to no fluid flow (Rudyk and Al-Lamki, 2015; Shanley et al., 2004). Variation in the height above the FWL that is derived from heterogeneity in rock properties has a significant effect on gas-water contact, and this parameter is in turn important for the calculation of gas-in-place, production, perforation, and hydraulic fracturing (Rudyk and Al-Lamki, 2015). In tight gas sandstone reservoirs that are often isolated, each reservoir has its FWL with no downdip water leg (Shanley et al., 2004).

9.6 Uncertainty analysis

The accuracy and validity of geological modeling depend mostly on a proper understanding of geological conditions, data used in modeling, and the rationality of the selected method and algorithm. The established model could be testified in various ways including geological knowledge validation, well pattern coarsening verification, reservoir parameter comparison, reserves calculation, and dynamic data verification. If the model accuracy was not verified by the mentioned ways, it should be re-established by using adjusted parameters (Zhi et al., 2015). In the tight sandstones, uncertainty with the results of the reservoir properties model is high. This is attributed to the large-scale and small-scale heterogeneity that controls the distribution of reservoir zones and pore system properties of reservoir rocks. Lithology (composition and texture) related to the depositional environment conditions, diagenetic processes, and structural framework are the main parameters that control reservoir heterogeneity. Therefore a detailed analysis of sedimentary facies and diagenesis, as well as the study of burial history and tectonic evolution of the tight gas are among the basic prerequisites that should be considered for reservoir modeling and interpretation of results. There are many tight sandstone reservoirs in which, despite the influence of diagenetic processes, facies and sedimentary environment have fundamental control over their reservoir characteristics and the distribution of production zones in the field. Tight sandstones of the Whicher Range field in Perth Basin of Western Australia, as mentioned above, are good examples in this regard. In these reservoirs, facies-based property modeling can efficiently unravel the reservoir heterogeneity and depict the distribution of reservoir zones. In contrast, in tight litharenite sandstones of the Moghan Basin in Iran, diagenesis has significantly modified reservoir properties, and it is difficult to establish a clear relationship between sedimentary facies and their reservoir properties. In these reservoirs where diagenesis plays the main role in reservoir properties, rock typing in the context of diagenetic facies and petrophysical rock types (e.g., electrofacies and flow units) can provide more reliable results of reservoir heterogeneity and establishing property modeling.

References

Al-Khalifah, M., Makkawi, M., 2002. The impact of data integration on geostatistical porosity modelling: a case study from the Berri field, Saudi Arabia. Journal of Petroleum Geology 25 (4), 485–498.

Ciftci, B., 2012. First order reservoir modeling of the Whicher Range Field tight gas sands. In: Whicher Range Tight Gas Sands Study, Western Australian Energy Research Alliance, Report 112, pp. 288–307.

Cuddy, S., 2017. Using fractals to determine a reservoir's hydrocarbon distribution. SPWLA 58th Annual Logging Symposium, pp. 17–21.

Izadi, M., Ghalambor, A., 2012. A new approach in permeability and hydraulic flow unit determination. SPE Reservoir Evaluation & Engineering 16 (3), 257–264.

Jamiolahmady, M., Sohrabi, M., Tafat, M., 2007. Estimation of saturation height function using capillary pressure by different approaches. SPE 107142.

Kadkhodaie, A., Kadkhodaie, R., 2022. Acoustic, density, and seismic attribute analysis to aid gas detection and delineation of reservoir properties, Chapter 3. In: Wood, D.A., Cai, J. (Eds.). In: The Fundamentals and Sustainable Advances in Natural Gas Science and Engineering, Sustainable Geoscience for Natural Gas Subsurface Systems, Gulf Professional Publishing 2, pp. 51–92.

Kadkhodaie, R., Kadkhodaie, A., Rezaee, R., 2021. Study of pore system properties of tight gas sandstones based on analysis of the seismically derived velocity deviation log: a case study from the Perth Basin of western Australia. Journal of Petroleum Science and Engineering 196, 108077.

Kadkhodaie-Ilkhchi, R., Kadkhodaie, A., Rezaee, R., Mehdipour, V., 2019. Unraveling the reservoir heterogeneity of the tight gas sandstones using the porosity conditioned facies modelling in the Whicher Range field, Perth Basin, Western Australia. Journal of Petroleum Science and Engineering 176, 97–115.

Kadkhodaie-Ilkhchi, R., Moussavi-Harami, R., Rezaee, R., Nabi-Bidhendi, M., Kadkhodaie-Ilkhchi, A., 2014. Seismic inversion and attributes analysis for reservoir quality evaluation of the tight gas sandstones of the Whicher Range field in the Perth Basin, Western Australia. Journal of Natural Gas Science and Engineering 21, 1073–1083.

Kadkhodaie-Ilkhchi, R., Rezaee, R., Moussavi-Harami, R., Kadkhodaie-Ilkhchi, A., 2013. Analysis of the reservoir electrofacies in the framework of hydraulic flow units in the Whicher Range Field, Perth Basin, Western Australia. Journal of Petroleum Science and Engineering 111, 106–120.

Lukoil Report, 2007. Geological model creation, delineation and estimation of prospects for Moghan and Lali blocks (Iran). In: Joint Study Project, NIOC Exploration/Luk Oil Company. Joint Study Project. Final Report to Contract No. IR 40-06-71-001-23/06.

Ma, Y.Z., Seto, A., Gomez, E., 2009. Depositional facies analysis and modeling of Judy Creek reef complex of the Late Devonian Swan Hills, Alberta, Canada. AAPG Bulletin 93 (9), 1235–1256. Available from: https://doi.org/10.1306/05220908103.

Moore, W.R., Ma, Y.Z., Pirie, I., Zhang, Y., 2016. Tight gas sandstone reservoirs, part 2: petrophysical analysis and reservoir modeling, Chapter 15. In: Unconventional Oil and Gas Resources Handbook, pp. 429–448.

Orsini, C., Rezaee, R., 2012. Depositional systems sequence stratigraphy frameworks & geological modelling of fluvial bodies. In: Geological Survey of Western Australia, Report 112, 405p.

Rudyk, S., Al-Lamki, A., 2015. Saturation-height model of Omani deep tight gas reservoir. Journal of Natural Gas Science and Engineering 27, 1821–1833.

Shanley, K.W., Cluff, R.M., Robinson, J.W., 2004. Factors controlling prolific gas production from low-permeability sandstone reservoirs: implications for resource assessment, prospect development, and risk analysis. AAPG Bulletin 88, 1083–1121.

Wei, W., Zhu, X., Meng, Y., Xiao, L., Xue, M., Wang, J., 2016. Porosity model and its application in tight gas sandstone reservoir in the southern part of west depression, Liaohe Basin, China. Journal of Petroleum Science and Engineering 141, 24–37.

Wu, H., Zhang, C., Ji, Y., Liu, R., Cao, S., Chen, S., et al., 2018. Pore throat characteristics of tight sandstones of Yanchang formation in eastern Gansu, Ordos Basin. Petroleum Research 3, 33–43.

Zhi, G., Longde, S., Ailin, J., Tao, L., 2015. 3-D geological modeling for tight sand gas reservoir of braided river facies. Petroleum Exploration and Development 42 (1), 83–91.

CHAPTER 10

Geomechanical properties of tight sandstones

10.1 Introduction

Deep tight gas sandstones with permeabilities less than 0.1 mD and effective porosities in the range of 5%–10% suffer from wide breakouts (BOs) and rotation of stresses along the borehole axis during drilling and completion. Increasing the flow rate in tight gas sandstones requires adopting appropriate well-stimulation technologies. Drilling horizontal wells and multi-stage fracturing have effectively been used to address the production issues relevant to the tight gas sandstone reservoirs. Optimization of perforation intervals for hydraulic fracturing jobs has become a major challenge in production from future wells. Therefore an accurate estimation of elastic rock properties, principle stresses magnitude and azimuth, field stress regime, and pore pressure plays a pivotal role in mechanical earth modeling (MEM) and addressing geomechanical issues of these reservoirs such as wellbore stability analysis and hydraulic fracturing design. This chapter presents the data and main steps required to make a geomechanical model, the main challenges, and the procedure to reduce the uncertainty in the geomechanical modeling of tight gas reservoirs.

10.2 Rock elastic properties

Rock properties of tight gas sandstones can be derived from core/cuttings (static) or well log data (dynamic). The most important petrophysical data used in dynamic data calculation are shear and compressional wave velocity together with density and gamma-ray logs. The main weight is on shear wave velocity which is not measured using conventional well log data. Normally, it is derived from an ultrasonic test on core plugs in the laboratory. The alternative method is advanced logging with Dipole Shear Sonic Imager (DSI) to measure both VP and Vs. Due to the high costs of measuring, a limited number of wells have DSI data. Accordingly, Vs estimation is an important issue for geomechanical modeling of tight gas sandstones.

Reservoir Characterization of Tight Gas Sandstones
DOI: https://doi.org/10.1016/B978-0-323-90180-2.00007-4

Normally, dynamic geomechanical data are used for the evaluation of formation strength using logging well data. In order to derive formation strength evaluation following parameters are used

1. Uniaxial Compressive Strength (UCS)
2. Friction angle
3. Cohesion
4. Young's modulus
5. Shear modulus
6. Bulk modulus
7. Poisson's ratio

Sources of strength evaluation are laboratory rock mechanics tests, petrophysical logs, and cuttings analysis. Normally, laboratory tests (static data) are a good and reliable source of rock strength measurement but have some deficiencies such as the limited amount of available core, time-consuming, costly and biased toward stronger intervals, and discontinuous. However, the petrophysical data are available as continuous logs for all intervals. So, after calibration with the static data, they are good sources of mechanical rock properties. The main mechanical rock properties derived from well log data are as follows.

Poisson's ratio is the ratio of transverse contraction strain to longitudinal extension strain in the direction of the stretching force. Tensile deformation is considered positive and compressive deformation is considered negative. The definition of Poisson's ratio contains a minus sign so that normal materials have a positive ratio.

Dynamic Poisson's ratio is calculated using shear (DTs in μs/f) and compressional (DTco in μs/f) logging data as follows.

$$\nu = \frac{0.5 \times (\mathrm{DTs}/\mathrm{DTco})^2 - 1}{(\mathrm{DTs}/\mathrm{DTco})^2 - 1} \quad (10.1)$$

In terms of velocity, Poisson's ratio is calculated as follows.

$$\nu = 0.5\left(\frac{Vp^2 - 2Vs^2}{Vp^2 - 2Vs^2}\right) \quad (10.2)$$

where Vp is calculated as 304.8/DTco (km/s) and $Vs = 304.8/\mathrm{DTs}$ (km/s).

Young's modulus, also known as the tensile modulus or elastic modulus, is a measure of the stiffness of an elastic material and is a quantity used to characterize materials. It is defined as the ratio of stress to strain. In solid mechanics, the slope of the stress-strain curve at any point is called the tangent modulus. The tangent modulus of the initial and linear portion of a stress-strain curve is called Young's modulus. It can experimentally be determined from the slope of a stress-strain curve created during tensile tests conducted on a sample of the material. In anisotropic materials, Young's modulus may have different values depending on the direction of the applied force with respect to the material's structure.

Young's modulus is the most common elastic modulus, sometimes called the modulus of elasticity, but there are also other measured elastic moduli such as the bulk modulus and the shear modulus. Young's modulus is calculated as follows.

$$E = \frac{\text{tensile} \quad \text{stress}}{\text{tensile} \quad \text{strain}} = \frac{\sigma}{\varepsilon} = \frac{F/A_0}{\Delta L/L_0} = \frac{FL_0}{A_0 \Delta L} \tag{10.3}$$

E is Young's modulus (modulus of elasticity)F is the force exerted on an object under tension;A_0 is the original cross-sectional area through which the force is applied;ΔL is the amount by which the length of the object changes;L_0 is the original length of the object.

Dynamic Young's modulus is calculated using density log (g/cc), shear (DTs in μs/ft), and compressional (DTco in μs/ft) logging data as follows

$$E(\text{psi}) = \frac{\rho_b}{\text{DTs}^2}\left(\frac{3\text{DTs}^2 - 4\text{DTco}^2}{DTs^2 - DTp^2}\right) \times 1.34 \times 10^{10} \tag{10.4}$$

The coefficient 1.34×10^{10} is used to convert Pa to psi.

Shear modulus or modulus of rigidity, denoted by G, or sometimes S or μ, is defined as the ratio of shear stress to the shear strain.

$$G = \frac{\tau_{xy}}{\gamma_{xy}} = \frac{F/A}{\Delta x/l} = \frac{Fl}{A\Delta x} \tag{10.5}$$

where

$$\tau_{xy} = F/A \tag{10.6}$$

F is the force that acts$A$ is the area on which the force acts and,

$$\gamma_{xy} = \frac{\Delta x}{l} = \tan\theta = \text{shear strain} \tag{10.7}$$

Shear Modulus is calculated from bulk density (g/cc) and shear transit time (μs/ft) using the following equation.

$$G(psi) = \frac{\rho_b}{DTs^2} \times 1.34 \times 10^{10} \tag{10.8}$$

The bulk modulus (K or B) of a rock measures the substance's resistance to uniform compression. It is defined as the ratio of the infinitesimal pressure increase to the resulting relative decrease of the volume. Its base unit is the Pascal. The bulk modulus $K > 0$ can formally be defined by the following equation.

$$K = -V\frac{dP}{dV} \tag{10.9}$$

where P is pressure, V is volume, and $\frac{dP}{dV}$ denotes the derivative of pressure with respect to volume. The inverse of the bulk modulus gives a rock's compressibility.

Bulk Modulus is calculated from bulk density (g/cc) and shear and compressional transit time (μs/ft) as follows.

$$K(psi) = \rho_b\left(\frac{1}{\mathrm{DTco}^2} - \frac{2}{3\mathrm{DTs}^2}\right) \times 1.34 \times 10^{10} \tag{10.10}$$

The ultimate (unconfined or uniaxial) compressive strength (UCS) of a material is the value of uniaxial compressive stress reached when the material fails completely. The compressive strength is usually obtained experimentally by means of a compressive test. The apparatus used for this experiment is the same as that used in a tensile test. However, rather than applying a uniaxial tensile load, a uniaxial compressive load is applied. As can be imagined, the specimen (usually cylindrical) is shortened as well as spread laterally. McNally (1987) proposed the following correlation for estimation of UCS in sandstones.

$$UCS_{MPa} = 1200 \times \exp(-0.036 \times DT) \tag{10.11}$$

10.2.1 Cohesion and friction angle

Cohesion(c) and internal friction angle (ϕ) are important rock properties for the construction of MEM and wellbore stability analysis. They represent the shear strength of rocks. Rocks with high internal friction angles have high cohesion and vice versa. In the case of shear failures, rocks are assumed to follow from Mohr-Coulomb failure criterion as follows.

$$\tau = c + \sigma' \tan\phi \tag{10.12}$$

where τ is shear stress, c is cohesion, ϕ is friction angle, and σ' is normal stress acting on the rupture surface.

Several correlations have been developed for the estimation of c and ϕ from well logging data. One of the most used ones for sandstones is cohesion estimation from the gamma-ray log as follows (Schlumberger Limited, 2015).

$$\phi = (-0.1166 \times GR) + 39.25 \quad (10.13)$$

To fall within the standard limits, the friction angles less than 15 degrees are set to 15 degrees, and those greater than 40 degrees are set to 40degrees.

Rock cohesion (c) can be obtained from UCS and friction angle as follows (Jaeger et al., 2007; Zhang and Yin, 2017).

$$c = \frac{UCS}{2 \times \sqrt{1 + (\tan\phi)^2} + \tan\phi} \quad (10.14)$$

Normally, the dynamic elastic rock properties overestimate the real rock strength and therefore the dynamic elastic properties should be calibrated to the static data derived from laboratory measurements.

10.3 Pore pressure in tight gas sandstone reservoirs

The pressure of fluids within the pore space of a reservoir is called pore pressure. In other words, it results from the pressure of a column of water from sea level to formation depth. In the homogeneous reservoirs, pore pressure is simply calculated from the weight of overlain fluids. In heterogeneous tight gas sandstones, the local changes in formations fluids and lithology, cause pore pressure to deviate from the hydrostatic pressure. Pore pressure in some tight gas reservoirs is at over-pressured levels and can even be equal to 95% of the vertical stress. Such abnormal pore pressures can increase drilling risks dramatically and cause serious drilling incidents, such as kicks, blowouts, and fluid influx if the pore pressure is not estimated accurately. Uncertain pore pressure estimation can lead to miscalculations in safe mud weigh window design that can lead to wellbore instability. Wellbore instability, shear, and compressive failure lead to BOs, pack-off, and casing collapse. Accordingly, mud loss and lost circulation can occur due to drilling-induced fractures caused by tensile failures. In worst situations, a borehole can be lost due to well instability. There are several approaches for pore pressure estimation including empirical equations from well logs and formation pressure tests such as RFT or MDT results. Eaton's (1975) method is one of the most popular approaches for formation pore pressure estimation in tight sandstones based on the sonic log variations.

$$P_p = P_{\text{ob}} - (P_{\text{ob}} - P_h) \times \left(\frac{\Delta T_n}{\Delta T_{\log}}\right)^x \tag{10.15}$$

where P_p is pore pressure, P_{ob} is overburden pressure, P_h is hydrostatic pressure, ΔT_n is the shale transit time of shale in normal compaction case, and $\Delta T_{\log}$ is sonic log reading. The exponent x of the equation is obtained by calibration of formation tests (e.g., MDT, DST). By default, the exponent x is taken as 3.

The normal compaction curve can be obtained by the equation proposed by Zhang (2011) as follows.

$$\Delta T_n = \Delta T_m + (\Delta T_{\text{ml}} - \Delta T_m) \times e^{-CZ} \tag{10.16}$$

where ΔT_n is the sonic transit time of shale (μs/f) in normal compactionΔT_m is sonic transit time in shale matrix (μs/f), and ΔT_{ml} is sonic transit time in ground level (μs/f), Z is depth (f) and C is a constant derived from curve fitting.

In addition to sonic log for pore pressure estimation, Eaton's method can be expressed based on resistivity logs or drilling d-exponent whose details of computation can be found in Rehm and McClendon (1971), Jorden and Shirley (1966), Babu and Sircar (2011). The estimated pore pressures and equation parameters need to be calibrated and optimized with pressure tests such as MDT.

10.4 Stress regime analysis

The Anderson scheme classifies an area as being characterized by normal, strike-slip, or reverse faulting depending on whether (1) the crust is extending and steeply dipping normal faults accommodate movement of the hanging wall (the block of rock above the fault) downward with respect to the footwall (the block below the fault), (2) blocks of crust are sliding horizontally past one another along nearly vertical strike-slip faults or (3) the crust is in compression and relatively shallow-dipping reverse faults are associated with the hanging wall block moving upward with respect to the footwall block. The Anderson classification scheme also defines the horizontal principal stress magnitudes with respect to the vertical stress. The vertical stress, S_V, is the maximum principal stress in normal faulting regimes, while the intermediate principal stress (SH) is the maximum principal stress in strike-slip and the reverse faulting regimes (Fig. 10–1 and Table 10.1).

Image logs such as resistivity-based full-bore formation micro imager (FMI) and acoustic-based Ultrasonic Borehole Imager (UBI) are a good source for calculation of main stresses direction. Using the interpretation of image logs in addition to the calculation of fracture parameters such as fracture density, and fracture aperture, their azimuth can also be obtained. Through analyzing a rose diagram of fractures azimuth, the direction of maximum horizontal stresses can be obtained. Drilling-induced fractures (IFs) and BOs observed in the image logs of the Lower Cretaceous

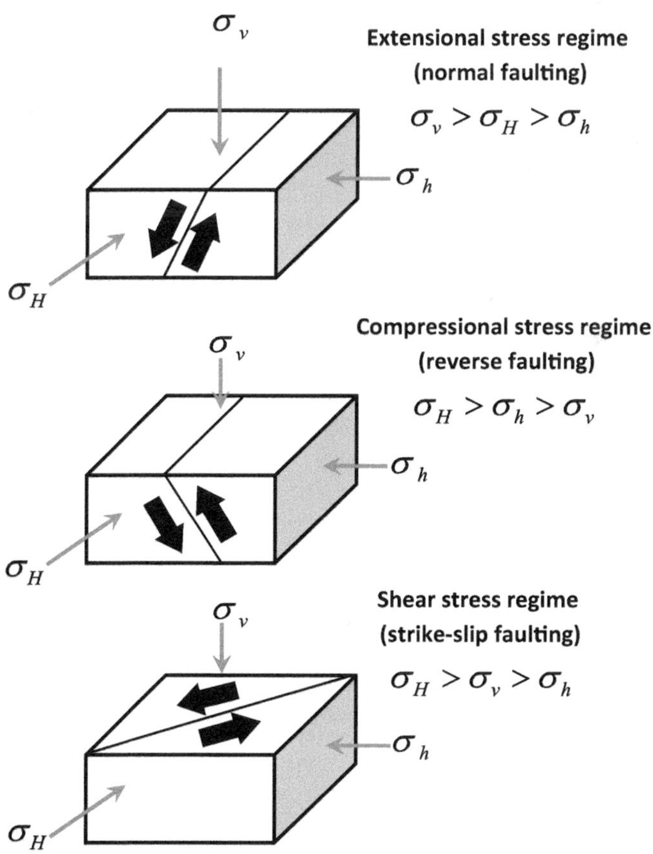

Figure 10–1 Anderson's classification scheme for relative stress magnitudes in normal, strike-slip, and reverse faulting.

Table 10.1 Relative stress magnitudes and faulting regimes.

Stress regime	S_1	S_2	S_3
Normal	S_v	S_{Hmax}	S_{hmin}
Strike-slip	S_{Hmax}	S_v	S_{hmin}
Revese	S_{Hmax}	S_{hmin}	S_v

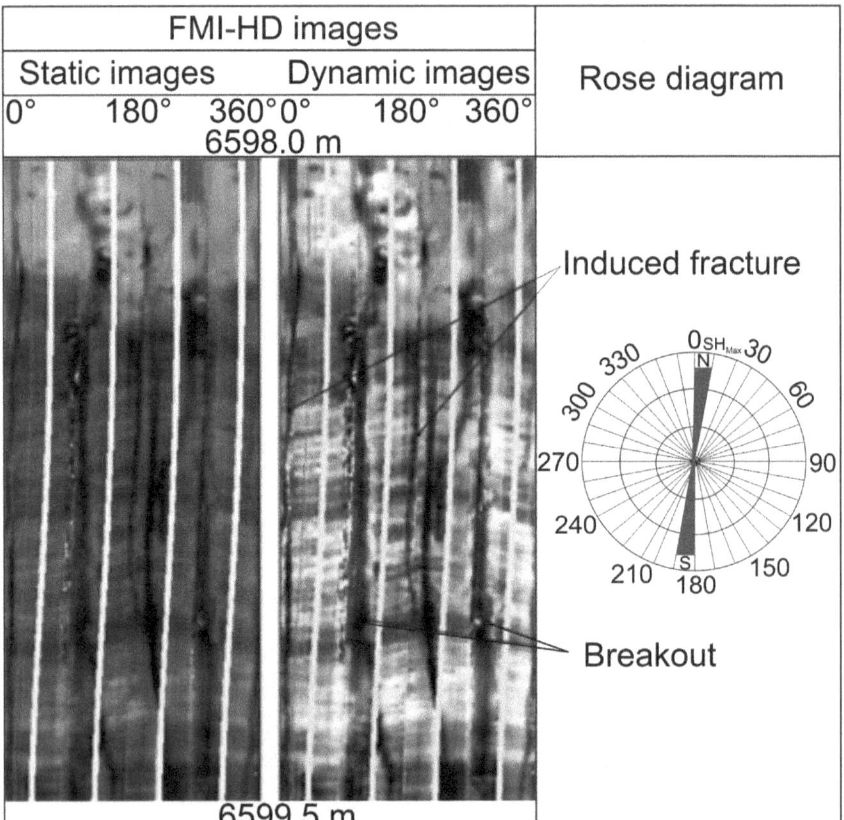

Figure 10–2 Drilling-induced fractures and breakouts observed in the Lower Cretaceous Bashijiqike tight gas sandstones of Kuqa depression, Tarim Basin. IFs are seen in a north-south trend, while BOs follow a west-east trend (Lai et al., 2017).

Bashijiqike tight gas sandstones are shown in Fig. 10–2. As is seen in Fig. 10–3, the induced fractures are seen in parallel to the direction of maximum horizontal stress (σ_H), while BOs are aligned to the direction of minimum horizontal stress (σ_h).

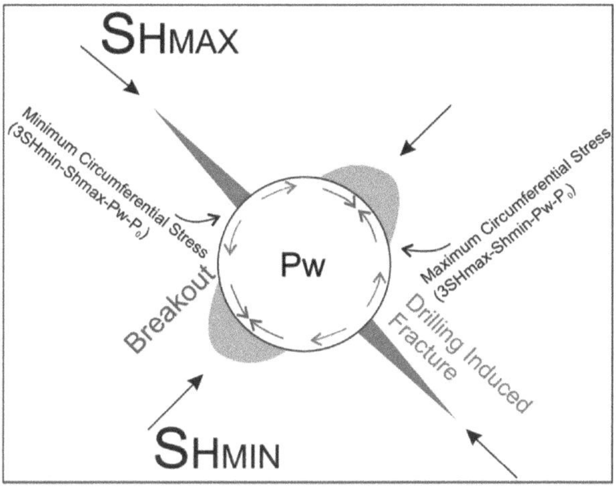

Figure 10–3 Drilling-induced fractures are parallel to maximum horizontal stress and breakouts are aligned with minimum horizontal stress direction (Fjaer et al., 1992).

10.5 Principal stresses calculation

10.5.1 Calculation of overburden pressure (vertical stress)

Vertical stress is an integration of hydrostatic and lithostatic pressures. As density log is a response of rock matrix and fluids it can be used for vertical stress calculation (S_v or σ_v) using the following equation. For this purpose, the density log is integrated with respect to depth.

$$S_v = g\left[\int_0^b \rho_{\text{sea}}.dz + \int_b^z \rho_b.dz\right] \tag{10.17}$$

For intervals where the density log is not available, the relationship between density and depth can be used to compute vertical stress. As an alternative, at the intervals for which no wire-line logs are run, an average value of the density log is considered based on the lithology derived from drilling data (e.g., graphic well logs).

10.5.2 Prediction of horizontal stress (σ_{Hmax} & σ_{hmin})

The magnitude of the maximum and minimum horizontal stress is an important input for geomechanical models. Calculation of horizontal stresses is always complex and needs some assumptions to be made. The poro-elastic theory is often used to estimate the minimum horizontal stress. Poro-elasticity is a continuum theory for the analysis of a porous media consisting of an elastic matrix containing interconnected fluid-saturated pores. In physical terms, the theory postulates that when a porous material is subjected to stress, the resulting matrix deformation leads to volumetric changes in the pores. Since the pores are fluid-filled, the presence of the fluid not only acts as a stiffener of the material but also results in the flow of the pore fluid (diffusion) between regions of higher and lower pore pressure. If the fluid is viscous the behavior of the material system becomes time-dependent. Using the following equations principal horizontal stresses can be calculated.

$$\sigma_h = \frac{\upsilon}{1-\upsilon}\sigma_V - \frac{\upsilon}{1-\upsilon}\alpha P_P + \alpha P_P + \frac{E}{1-\upsilon^2}\varepsilon_x + \frac{\upsilon E}{1-\upsilon^2}\varepsilon_y \tag{10.18}$$

$$\sigma_H = \frac{\upsilon}{1-\upsilon}\sigma_V - \frac{\upsilon}{1-\upsilon}\alpha P_P + \alpha P_P + \frac{\upsilon E}{1-\upsilon^2}\varepsilon_x + \frac{E}{1-\upsilon^2}\varepsilon_y \tag{10.19}$$

where α is the Biot's constant (usually assumed 1), P_p is pore pressure, ν is Poisson's ratio, E is the Young modulus, while ε_x and ε_y are the elongation components of the linear strain on the "Y" and "X" axis. ε_x and ε_y parameters can be calculated from the leak-off test (LOT) or extended leak-off test (XLOT) results. In the absence of LOT and XLOT, the elongation components, are optimized using iterative numerical solution methods in a way that the main stresses configuration met the faulting system (stress regime).

It is worth mentioning that the mini-frac test or LOT can be used to derive the minimum horizontal stress and to calibrate the results of σ_h estimation.

10.6 Geomechanical modeling

Geomechanical models of tight gas sandstones address the issues relevant to wellbore stability in underbalanced and near balanced drilling (designing the safe mud window and collapse pressure), hydraulic fracturing design, sand production, and casing collapse. The borehole stratigraphic information including the drilled formations from surface to bottom-hole, MEM, and an appropriate failure criterion is employed to make a geomechanical model. A MEM includes rock elastic properties, UCS, principal stresses, and pore pressure. An appropriate failure criterion is applied to the geomechanical model to derive threshold pressures such as collapse pressure, fracture initiation pressure (FIP), and formation breakdown pressure (FBP). Various failure criteria are proposed by Drucker-Prager (1952), Wiebols and Cook (1968), Lade (1977), Hoek-Brown (1980), and Mogi-Coulomb (Al-Ajmi and Zimmerman, 2005, 2006). A general workflow to make a geomechanical model is shown in Fig. 10–4.

10.7 Hydraulic fracturing

The commercial production from tight gas sandstones is heavily dependent on the hydraulic fracturing operation. Through hydraulic fracturing, new fractures are induced in a tight gas formation to create a supplementary fracture network and enhance flow rates (Xu et al., 2009; Kresse et al., 2013; Wang, 2019). In this section, different aspects to execute a hydraulic fracturing job in the tight gas sandstones include MEM design, selection of candidate wells/intervals, selection of propping agent, field execution of hydraulic fracturing, fracture propagation monitoring, and fracture modeling will be discussed in brief.

10.7.1 Mechanical earth model

Prior to any hydraulic fracturing job, a geomechanical study is required to design the operation parameters and choose the candidate wells/intervals. Using an appropriate MEM, the candidate intervals for hydraulic fracturing, FIP, FBP, and fracture propagating pressure are estimated. Fracture opening is parallel to the direction of minimum horizontal stress (σ_h), while their propagation is perpendicular to the azimuth σ_h. Accordingly, in the reverse faulting stress regime (compressional stress system), the hydraulic fractures (HF) are horizontal, while in the normal faulting stress regime (extensional stress system) vertical fractures will propagate (Fig. 10–5). As mentioned earlier in this chapter, mechanical rock properties especially in-situ stresses, Poisson's ratio, and Young's modulus are the important parameters to optimize the completion. The Young's modulus reflects the stiffness of formation and controls the width of fracture (fracture opening). Poisson's ratio as a measure of lateral deformation of formation in response to stress is required in many fracturing design equations.

FIP: In order to initiate hydraulic fracturing in the interval of interest, high-pressure treatment is injected into the wellbore to create tensile HF. In this regard, in-situ stress is the most important rock property controlling the pressure required for the creation and propagation of fractures within the formation. It is often called fracture closure pressure (FCP) or minimum compressive strength that is required to keep the hydraulic

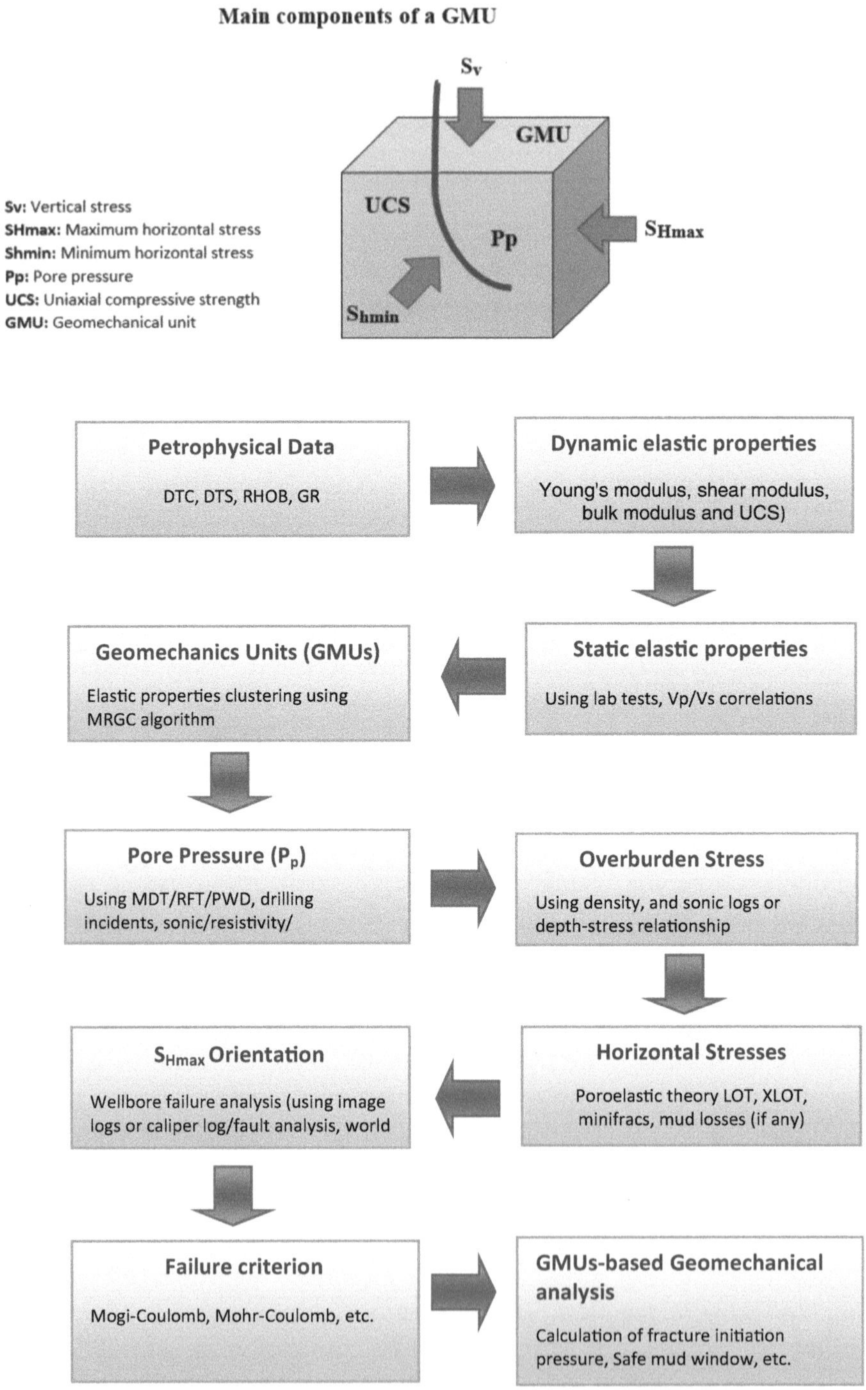

Figure 10–4 A general workflow to make a geomechanical model.

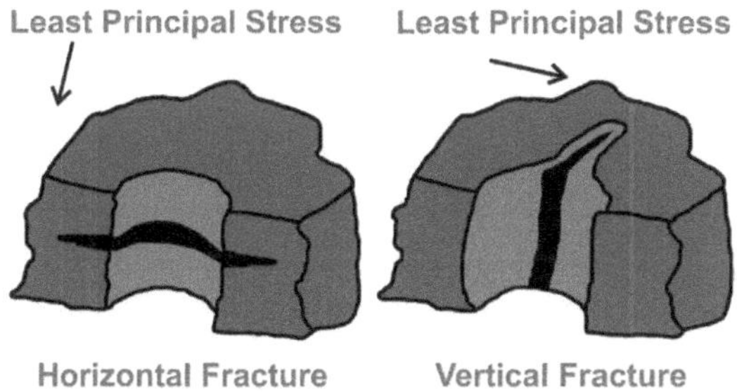

Figure 10–5 Fracture propagation and opening with regard to the minimum horizontal stress. Fracture opening is parallel to minimum horizontal stress, while its propagation is along with the maximum horizontal stress. In the reverse fault system, where vertical stress is the minimum stress ($S_{Hmax} > S_h > Sv$), fractures are horizontal planes, while in the normal fault system fractures are vertical planes.

fracture open. In order to keep the fracture open, the pressure inside the fracture should be greater than the minimum compressive strength. In-situ stress can be obtained from well logs, core tests, or injection tests.

FBP: The FBP is one of the important parameters that need to be estimated prior to hydraulic fracturing. After initiation of a hydraulic fracture, additional fluid injection is still required to reach the FBP. FBP has an upper and lower limit. The lower limit of FBP is equal to minimum horizontal stress (Sh_{min}) or σ_h. The upper limit of FBP is calculated from the Eq. (10.20) as follows (Haimson and Fairhurst, 1970; Zoback, 2007; Zhang and Roegiers, 2010).

$$P_b = 3\sigma_h - \sigma_H - P_p - \sigma_T + T_0 \tag{10.20}$$

where P_b is the upper limit of FBP, σ_h is minimum horizontal stress, σ_H is maximum horizontal stress, P_p is pore pressure, σ_T is thermal stress due to the temperature difference between formation and drilling mud, and T_0 is formation tensile strength.

If the value of tensile strength and thermal stress is negligible, then P_b is simplified as follows.

$$P_b = 2\sigma_h - P_p \tag{10.21}$$

As discussed in Sections 10–5.2, σ_h can be derived from poroelastic theory as follows.

$$\sigma_h = \frac{v}{1-v}\left(\sigma_v - P_p\right) + P_p \tag{10.22}$$

Replacing σ_h Eq. (10.21) with Eq. (10.22), the following equation is derived for FBP calculation.

$$P_b = \frac{2\upsilon}{1-\upsilon}\left(\sigma_v - P_p\right) + P_p \tag{10.23}$$

Accordingly, the mean P_b (average of the upper and lower limit), as an important parameter in hydraulic fracturing job design, is derived from Eq. (10.24).

$$P_b = \frac{3\upsilon}{2(1-\upsilon)}\left(\sigma_v - P_p\right) + P_p \tag{10.24}$$

For a horizontal well drilled along with the minimum horizontal stress, the maximum breakdown pressure can be calculated using the following equations.

$$P_{bH} = 3\sigma_H - \sigma_V - P_p + kT_0,\ if\sigma_V > \sigma_H \tag{10.25}$$

$$P_{bH} = 3\sigma_H - \sigma_V - P_p + kT_0,\ if\sigma_V \leq \sigma_H \tag{10.26}$$

In order to calibrate and determine of *k* value, fracture analysis tests such as XLOT and diagnostic fracture injection tests (DFIT) are utilized. If fracture initiation and formation breakdown happen at the same time, then *k* is taken as 1, otherwise $k > 1$. The default value of *k* is $\sqrt{2}$. DFIS also called mini fall-off or mini-frac test is an indirect technique to estimate FCP in tight gas sandstones. DFIT is a small-scale fracturing test where small quantities of a clear fluid (e.g., 2% or 3% KCl) are injected to create a fracture before pressure release and the well-head is shut in (Zhang et al., 2018).

Fracture propagation pressure (FPP): After reaching the formation breakdown pressure, the fluid pressure inside the hydraulic fracture should be high enough to overcome the minimum horizontal stress and propagate the tensile fracture through the formation. As proposed by Hubbert and Willis (1957), the injected fluid pressure should be equal to or greater than the minimum horizontal stress to keep the fracture open and extend it. Racture propagation pressure (FPP) ($P_{b\text{min}}$) can be obtained from the following equation.

$$P_{b\text{min}} = \sigma_h + kT_0 \tag{10 - 27}$$

where *k* is a constant and usually is by default taken as $\sqrt{2}$. If fracture initiation and formation breakdown occur simultaneously, then $k = 1$.

Eq. (10.27) implies that pressure inside a fracture should overcome minimum horizontal stress and tensile strength of formation for fracture propagation (Zhang et al., 2018).

10.7.1.1 Leak-off test

LOT, also called the mini-frac test, is an important test for the determination of minimum horizontal stress. A typical diagram of a *LOT* is shown in Fig. 10–6. As is seen, the pressure increases as the volume of the pumped fluid increases until a distinct break in the curve slope or deviation from the linearity is reached which is called the Leak-off point (*LOP*). Alternatively, *LOP* is named P_i or FIP. The slope of the curve before the *LOP* or yielding point indicates Young's modulus. If the *LOP* is not reached, the test is called a formation integrity test (*FIT*) or limit

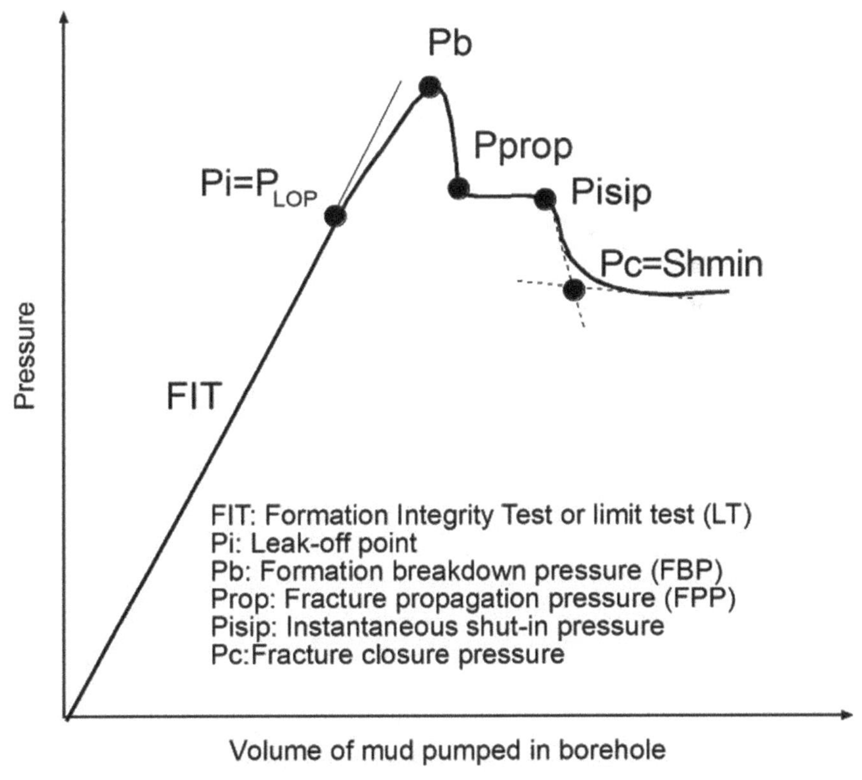

Figure 10–6 Diagram showing the pressure versus pumped fluid volume for LOT or mini-frac test. *LOT*, Leak-off.

test. The peak of the pressure-pumped volume indicates the formation breakdown pressure (*FBP*) or P_b. The stable growth of fracture happens between *LOP* and *FBP*. The continued pumping of the fracturing fluid into the candidate well after the peak point is associated with a drop in pressure. This is called FPP and indicates the unstable growth of the fracture. After passing the fracture propagation pressure, the fluid pumping into the well stops, and a sudden drop in pressure is observed which is called instantaneous shut-in pressure (*ISIP*). The last point that is important shows the FCP or P_c. It shows the pressure at which the newly created hydraulic fracture is closed and indicates the magnitude of minimum horizontal stress. Each of the *LOP*, *ISIP*, or *FCP* can be regarded as σ_h. White et al. (2002) consider *ISIP* as minimum horizontal stress while many other researchers such as Yamamoto (2003) take *FCP* as a representative value σ_h. To terminate the *LOT*, the rig floor valve opens and some part of the pumped fluid flows back to the fluid reservoir (bleed-off).

10.7.1.2 Extended leak-off test

The extended leak-off test (*XLOT*) is a longer and more advanced form of *LOT* and consists of two or more complete cycles. The first cycle is as with the *LOT* and consists of *LOP*, *FBP*, *FPP*, *ISIP*, and *FCP*. In the second cycle, all steps are repeated until unstable growth of fracture (second instantaneous shut-in pressure). In addition, the difference between the pressure peaks in the first cycle and the second cycle indicates the tensile strength of the formation (T_0). A representative diagram of *XLOT* is shown in Fig. 10–7.

10.7.2 Selection of candidate intervals and designing hydraulic fracturing job

Most of the tight gas sandstones are thick and layered reservoirs and choosing an optimal candidate well or interval controls the successful implementation of the fracturing job. Many variables should be considered for a successful hydraulic fracture treatment which can be mentioned FCP, formation permeability, reservoir fluid viscosity, formation pressure and depth, wellbore conditions, and skin factor. The best candidate wells for hydraulic fracturing in tight gas sandstones should have a thick pay zone with substantial gas initially in place (GIIP), large areal extent, medium to high pressure, and efficient in-situ stress barriers to suppress the vertical growth of fractures in the upper and lower interval of the pay

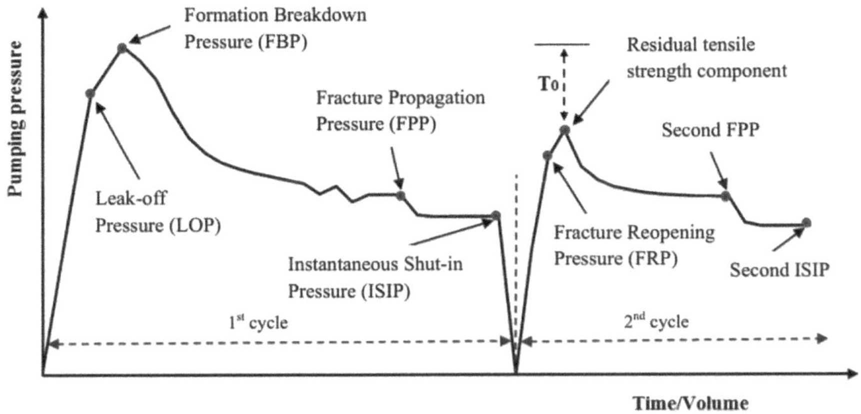

Figure 10–7 Diagram showing the pressure versus pumped fluid volume for XLOT. *XLOT*, Extended leak-off. After Zhang et al. (2017).

zone. That is, the candidate pay zones should be sandwiched between upper and lower stress barriers such as compacted clean shales. Thus low-pressure tight sandstones with a small areal extent and small volume of GIIP are considered poor candidates for hydraulic fracturing jobs. It should be noted that ultra-low permeability tight sandstones cannot produce economically to cover the drilling and completion expenses even if a successful hydraulic fracturing job is performed. During the design process, the optimum fracture length and fracture conductivity on the ultimate recovery is determined at minimum cost. Therefore a pre-job knowledge of elastic properties, in-situ stresses, and the presence of natural fractures (NF) is very important to estimating the fracture distribution patterns. Using the available data, a base case is run to determine which parameters are the most uncertain. Most of the time, parameters such as permeability, Young's modulus, FCP, and fluid loss coefficients are uncertain and the design engineer makes sensitivity analysis to determine their influence on the fracturing job and find the optimum values. The hydraulic fracturing model will also tell us the type of fracturing fluid to be pumped into well such that the best values of propped fracture length and fracture conductivity are obtained.

10.7.3 Selection of fracturing fluid and propping agent

The selection of an appropriate fracking fluid that effectively interacts with the hydraulic fracture wall is of prime importance. Each reservoir has a specific response to a particular type of fracking fluid that is assessed by performing a range of Diagnostic Fracturing Injection Tests (DFITs). Moreover, selecting the type of propping agent for hydraulic fracturing depends on the magnitude of maximum effective stress that needs to be applied to the propping agent.

The minimum flowing bottom-hole pressure controls the magnitude of maximum effective stress during the life of the well. Sand is used as a propping agent in case of having maximum effective stress of less than 6000 psi. In the case of having the maximum effective stress between 6000 and 12,000 psi, a medium-strength proppant such as resin-coated sands is employed. High strength proppant such as bauxite is used if the maximum effective stress is greater than 12,000 psi. Of course, some exceptions such as proppant flow back issue, availability of sand, and shipping costs should be considered (Economides and Nolte, 2000). The optimal fracturing fluid is lightweight and strong enough and it should not cause formation damage as gas-based (e.g., carbon dioxide and nitrogen) and water-based (e.g., gels and slickwater) muds. Spherical proppant perform better than angular ones to keep the fracture open such that gas can flow into the wellbore in economic volumes (Shehata et al., 2010). After determination of fracture half-length, fracturing fluid, and propping agent, the design engineer employs a P3D model to determine the optimum values of injection rate, pad volume, fluid loss additives, perforation locations, etc.

10.7.4 Field execution of hydraulic fracturing

Once the hydraulic fracturing design is optimized, the fracture treatment must be pumped into the candidate well. Hydraulic fracturing operation is a teamwork job and demands good cooperation and coordination between all parties. The safety team monitors the maximum values of allowable pressure, injection rate, and annulus pressure. Equipment and electronic devices should resist high pressure and high-pressure borehole environment during the treatment. Water samples as the base fracturing fluid need to be analyzed for the presence of minerals and bacteria types. These data will help in the selection of the appropriate additives to propagate a wider fracture and to transfer the proppant into the fracture. Using the technology of propellant-perforating stimulation enhances the efficiency of the fracturing job. A propellant is a mixture of fuel and an oxidizer that can create a huge amount of pressure when ignited due to generating a large volume of gases. The Casing Conveyed Perforating

System (CCPS) makes it possible to perform multilayer fracturing in a short time and improves the gas recovery (Shehata et al., 2010). The post-fracturing reservoir pressure and production data must be analyzed using an appropriate method that is consistent with the flow regimes occurring at the time of data collection (McGuire and Sikora, 1960; Prats, 1961; Gidley et al., 1989). Multi-hydraulic fracturing stage can enhance the production and flow rate. In thick tight gas formations, fracturing a horizontal well can increase production 2 to 3 times more than a vertical well. It is worth mentioning that increasing the propped length of fractures and proppant volume increases the production rate and revenue. Propped fracture half-length refers to the fracture length that will be supported by proppant after the closure of fractures. However, several factors can impose limitations on the size of the fracturing job and hence the propped length of HF such as the costs of the treatment (e.g., the costs of proppant and fluid), available pumping rates, and the investment policy of the operating company (Holditch et al., 1978; Veatch, 1983; Britt, 1985).

10.7.5 Fracture propagation monitoring

In order to monitor the propagation of the hydraulically created fractures, the microseismic technique is employed. This will help in measuring the length, azimuth, and opening of the hydraulic fracture (Maxwell et al., 2010). Microseismic technology was originally used in the mining industry and subsurface geo-energy to avoid geohazards. The main difference between conventional seismology and microseismic is the source of seismic waves. In conventional seismology, artificially induced sources are employed to create seismic waves (e.g., explosives, vibrators), while the microseismic method uses passive seismic technology (Duncan and Eisner, 2010). That is, in microseismic operation, weak seismic waves originating from the subsurface stresses change due to hydraulic fracturing operation are utilized. Microseismic technology is similar to earthquake seismology from the standpoint of seismic wave sources as both use passive seismic technology (Li et al., 2022). A schematic diagram of the microseismic survey for hydraulic fracturing monitoring is shown in Fig. 10–8. Using the robust computer imagery, the progressive fracture growth in 3D space and formation response to pumping pressure variations is monitored. This way, the fracture simulation design is optimized to maximize the production from fractures.

A typical microseismic survey to monitor the hydraulic fracturing in the tight gas sandstones of the Lower Permian Lower Shihezi Formation

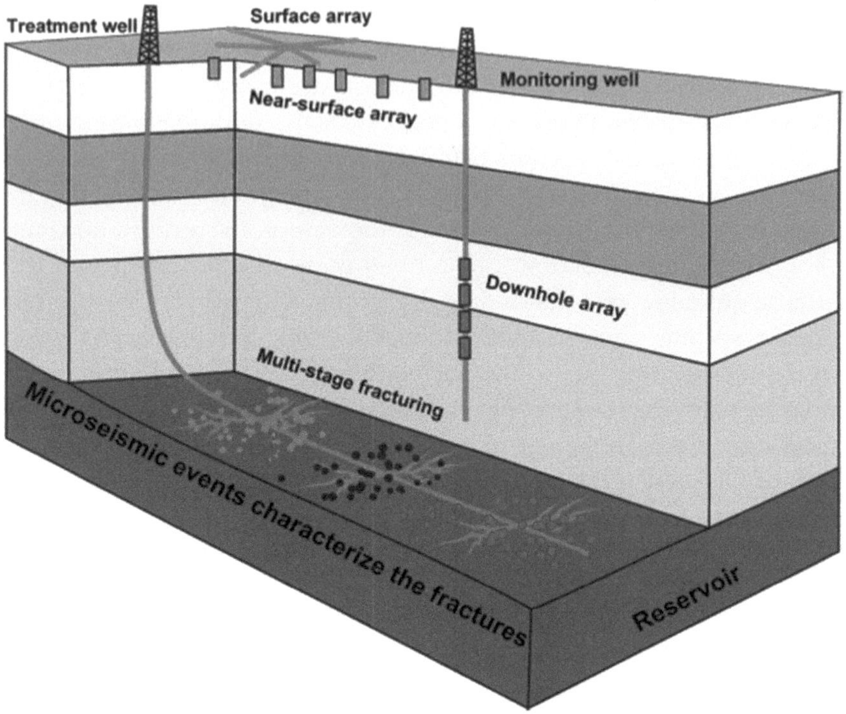

Figure 10–8 Diagram showing microseismic survey for hydraulic fracturing monitoring (Li et al., 2022).

of Daniudi Gas Field is shown in Fig. 10–9. As is seen, a relatively large amplitude caused by HF is generated at 3461 s. The results of microseismic monitoring in the first stage of hydraulic fracturing of well HH1 are shown in Fig. 10–10. The red and blue histograms represent fractures propagated in NE and SW directions, respectively.

10.7.6 Hydraulic fracture and natural fracture interaction

In a tight gas sandstone, HF can cross or open NF. When a pressurized hydraulic fracture (HF) intersects an unpressurized natural fracture (NF), one of the following scenarios may happen (Chuprakov et al., 2010; Gu et al., 2012; Wu and Olson, 2014; Wang, 2019; Weng et al., 2019).

1. HF will cross the NF if the difference between the minimum and maximum horizontal stress is high and fracturing fluid is associated with low leak-off and spur rates (e.g., linear gel).

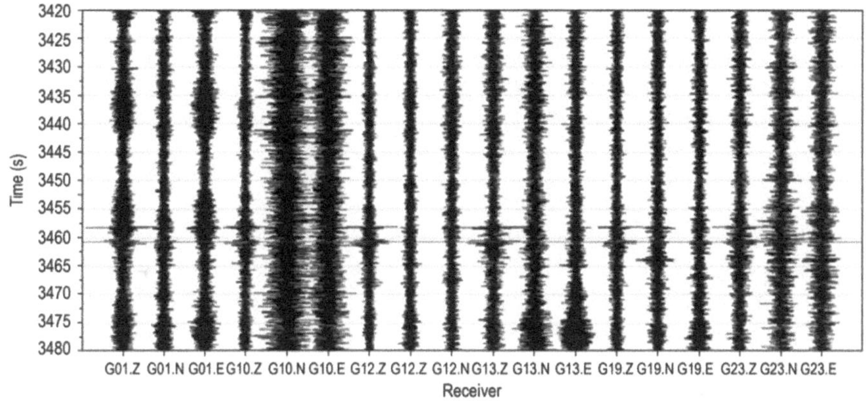

Figure 10–9 A typical microseismic survey indicating a relatively large amplitude at 3461 s caused by hydraulic fracturing in the tight gas sandstones of the Lower Permian Lower Shihezi Formation, Daniudi Gas Field (Rui-Zhao et al., 2013).

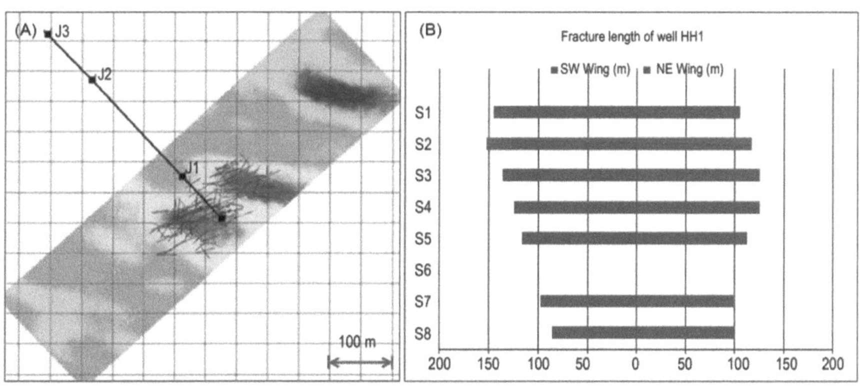

Figure 10–10 (A) The results of microseismic monitoring in the first stage of hydraulic fracturing of well HH1. The target reservoir is the tight gas sandstones of the Lower Permian Lower Shihezi Formation, Daniudi Gas Field; (B) Histogram showing the length of hydraulic fracture in the different fracturing stages. The hydraulic fractures propagated in NE and SW directions (Rui-Zhao et al., 2013).

2. HF will deviate from the NF if the difference between the minimum and maximum horizontal stress is moderate and the leak-off rate is low to moderate (e.g., linear gel or high injection rate of slickwater).
3. HF will be arrested by NF if the difference between the minimum and maximum horizontal stress is low and the leak-off rate is low to moderate and fracturing fluid is associated with high leak-off and spur rates (e.g., the low injection rate of water).

Other scenarios such as the formation of multiple fractures and diversion can be observed depending on the rock's elastic properties, stresses anisotropy, and fracturing fluid properties (Jeffrey et al., 2009). Stress anisotropy, which is defined as the difference between the stresses normal and parallel to the hydraulic fracture plane, is the main controlling factor for hydraulic fracture propagation.

The ability of hydraulic fracture to open NF depends on in-situ stresses, treatment pressure, elastic properties, and natural fracture properties (Zhang et al., 2018). As shown by Nolte and Smith (1981) and Nolte (1991), a hydraulic fracture intersecting at any arbitrary angle will open a natural fracture when the critical net pressure (P_{nf}) is achieved.

$$P_{\mathrm{nf}} = \frac{\sigma_H - \sigma_h}{1 - 2v} \tag{10.28}$$

This implies the net pressure in the fractures (P_n) should be greater than the critical net pressure (P_{nf}) to open the natural fracture ($P_n > P_{\mathrm{nf}}$). The net pressure in the fracture is defined as the difference between minimum horizontal stress and treatment pressure ($P_n = P_t - \sigma_H$). If treatment pressure is less than critical net pressure ($P_t < P_{\mathrm{nf}}$), then the hydraulic fracture will cross the natural fracture without making it open. The higher treatment pressure makes it easier to open the natural fracture. As it can be understood from Eq. (10.28), at the same treatment pressure P_t, a smaller difference between the maximum and minimum horizontal stresses makes it easier to open the NF (Zhang et al., 2018).

10.7.7 Molding of hydraulic fractures

After stimulation of the tight sandstones through hydraulic fracturing, the acquired micro-seismic data are processed and interpreted to monitor the fracture network growth. The results of interpretation of microseismic data integrated with the available geological, geomechanical, and reservoir data have led to the development of fracture network simulation models. Of course, due to the time and budget considerations, it is not always feasible to acquire micro-seismic data on the actual fracturing job. Therefore simulation of fracture parameters is a pivotal step in the production and development of tight gas sandstones. There are several algorithms to simulate the fracture network in tight gas sandstone with the pre-existing NF which can be mentioned Perkins-Kern-Nordgren (PKN) model, Khristianovic-Geertsma de Klerk (KGD) model, and Unconventional Fracture Model (UFM).

UFM: Unconventional Fracture Model (UFM) is a fully coupled numerical modeling algorithm to simulate the hydraulic fracture propagation and its relevant fluid flow (Wang, 2019, Kresse et al., 2013). UFM accounts for stresses anisotropy, reservoir heterogeneity, and 3D stress-shadow effects. It is similar to pseudo-3D fracture modeling algorithms and it is well suited to deal with HFs intersecting NFs at any angle of approach. The OpenT algorithm as the basis of UFM is used to solve the elasticity problem of HF and NF interaction and the condition for re-initiation of hydraulic fracture at natural fracture. The average width of hydraulic fracture ($\overline{\omega}$) is calculated as follows (Valko and Economides, 1995; Chuprakov et al., 2013).

$$\overline{w} = 2.53\left[\frac{Q\mu L^2}{H(E_s/(1-v_s^2))}\right] \tag{10.29}$$

where Q is injection volume of fracturing fluid, μ is the viscosity of fracturing fluid, E_s is static Young's modulus, v_s is static Poisson's ratio, H and L denote height and length of hydraulic fracture.

Upon interaction of a pressurized HF with an unpressurized NF, the pressure inside the natural fracture starts to increase by time after treatment commencement.

PKN model: PKN model was proposed by Perkins and Kern (1961) and Nordgren (1972) as a 2D analytical solution to estimate hydraulic fracture width. They assume that plane strain deformation is in the vertical plane. In this approach, fracture height is not dependent on the fracture length and it implies that fracture length is much greater than its height (Fig. 10–11). Perkins and Kern (1961) assume that HF is restricted to the pay zone. That is when the height of an HF is greater than the thickness of the pay zone, a hydraulic fracture will no longer propagate in the vertical direction and it will choose the horizontal plane at the top and bottom of the pay zone. The maximum width of the hydraulic fracture is estimated as follows.

$$w_{\max} = \frac{2(1-v^2)}{E}(P_t - \sigma_H)H \tag{10.30}$$

where $w_{\max}$ is the maximum width of fracture, P_t is treatment pressure, σ_H is maximum horizontal stress, H is fracture height, v is Poisson's ratio and E is Young's modulus.

The PNK model is not applicable when the ratio of fracture height to length is large. It is well suited for very long fractures. Zhang et al. (2018)

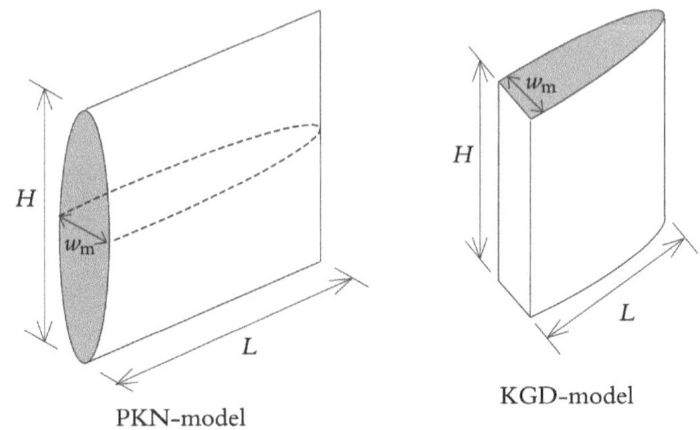

Figure 10–11 Illustration of PKN and KGD fracture models. The PKN model implies that fracture length is larger than fracture height, while in the KDG model fracture height is larger than fracture length (Fjaer et al., 2021).

proposed the following equation to estimate the maximum width of PKN fracture in a 3D state ($w_{3D\max}$).

$$w_{3D\max} = \frac{2(1-v^2)}{(1+H/L^{1.22})E}(P_t - \sigma_h)H \tag{10.31}$$

In order to estimate the length of the bi-wing planar fracture, the mass balance equation can be employed. That is, the volume of injected fracturing fluid and proppants should be equal to the volume of HF.

$$aV = L_{tot}Hw_{avg} \tag{10.32}$$

where V is the total volume of injected fracturing fluid and proppants, a is the fluid efficiency coefficient, L_{tot} is the total length of fracture, H is the fracture height, w_{avg} is the average width of fracture ($w_{avg} \approx 0.628w_{mac}$, w_{avg} can be estimated from Eqs. (10.30) to (10.31)] (Zhang et al., 2018).

KGD model: Assuming that fracture height is constant and much larger than fracture length (Fig. 10–11), Geertsma and de Klerk (1969) derived the following equation for estimation of maximum fracture width in a 2D plane (Sneddon, 1946).

$$w_{\max} = \frac{4(1-v^2)}{E}L(P_t - \sigma_h) \tag{10.34}$$

where $w_{\max}$ is maximum fracture width, P_t is treatment pressure, σ_h is minimum horizontal stress, L is fracture length, v is Poisson's ratio and E is Young's modulus.

References

Al-Ajmi, A.M., Zimmerman, R.W., 2005. Relationship between the parameters of the Mogi and Coulomb failure criterion. International Journal of Rock Mechanics & Mining Sciences 42 (3), 432–439.

Al-Ajmi, A.M., Zimmerman, R.W., 2006. Stability analysis of vertical boreholes using the Mogi-Coulomb failure criterion. International Journal of Rock Mechanics & Mining Sciences 43 (8), 1200–1211.

Babu, S., Sircar, A., 2011. A comparative study of predicted and actual pore pressure in Tripura, India. Journal of Petroleum Technology and Alternative Fuels 2 (9), 150–160.

Britt, L.K., 1985. Optimized oilwell fracturing of moderate-permeability reservoirs. Presented at the SPE Annual Technical Conference and Exhibition, Las Vegas, Nevada, USA, 22–26 September. SPE-14371-MS.

Chuprakov, D.S., Akulich, A.V., Siebrits, E., Thiercelin, M., 2010. Hydraulic fracture propagation in a naturally fractured reservoir. SPE 128715, Presented at the SPE Oil and Gas India Conference and Exhibition held in Mumbai, India, 20–22, January.

Chuprakov, D.A., Melchaeva, O.Y., Prioul, R., 2013. Hydraulic fracture propagation across a weak discontinuity controlled by fluid injection. Paper presented at the ISRM International Conference for Effective and Sustainable Hydraulic Fracturing, May 20–22, Paper Number: ISRM-ICHF-2013-008. International Society for Rock Mechanics and Rock Engineering.

Drucker, D.C., Prager, W., 1952. Soil mechanics and plastic analysis of limit design. Quarterly of Applied Mathematics 10, 157–165.

Duncan, P.M., Eisner, L., 2010. Reservoir characterization using surface microseismic monitoring. Geophysics 75, 139–146. 75A.

Eaton, B.A., 1975. The equation for geopressure prediction from well logs. Society of Petroleum Engineers of AIME. Paper SPE 5544.

Economides, M.J., Nolte, K.G., 2000. Reservoir Stimulation, third ed. John Wiley & Sons, Ltd., West Sussex, England, p. 856.

Fjaer, E., Holt, R.M., Horsrud, P., Raaen, A.M., Risnes, R., 2021. Chapter 11 – Mechanics of hydraulic fracturing, Developments in Petroleum Science, Vol. 72. Elsevier, pp. 555–600.

Fjaer, E., Holt, R.M., Horsrud, P., Raen, A.M., 1992. Petroleum Related Rock Mechanics, 2nd (Ed.) 515p.

Geertsma, J., de Klerk, F., 1969. A rapid method of predicting width and extent of hydraulically induced fractures. Journal.of Petroleum Technology 21 (12), 1571–1581. SPE-2458.

Gidley, J.L., Holditch, S.A., Nierode, D.E., Veatch Jr., R.W., 1989. Recent Advances in Hydraulic Fracturing, 12. Monograph Series, SPE, Richardson, TX, p. 317.

Gu, H., Weng, X., Lund, J., Mack, M., Ganguly, U., Suarez-Rivera, R., 2012. Hydraulic fracture crossing natural fracture at nonorthogonal angles: a criterion and its validation. SPE Production & Operations 27 (1), 20–26. SPE-139984.

Haimson, B.C., Fairhurst, C., 1970. In situ stress determination at great depth by means of hydraulic fracturing. In: Somerton, W.H. (Ed.), Rock Mechanics – Theory and Practice. The American Institute of Mining, Metallurgical, and Petroleum Engineers (AIME), pp. 559–584.

Hoek, E., Brown, E.T., 1980. Empirical strength criterion for rock masses. Journal of the Geotechnical Engineering Division Research 106, 1013–1035.

Holditch, S.A., Jennings, J.W., Neuse, S.H., Wyman, R.E., 1978. The optimization of well spacing and fracture length in low permeability gas reservoirs. Presented at the SPE Annual Fall Technical Conference and Exhibition, Houston, Texas, 1–3 October 1978. SPE-7496-MS.

Hubbert, M.K., Willis, D.G., 1957. Mechanics of hydraulic fracturing. Petroleum Transportation AIME 210, 153–168.
Jaeger, J.C., Cook, N.G., Zimmerman, R., 2007. Fundamentals of rock mechanics. John Wiley and Sons, Chicago, IL, p. 488.
Jeffrey, R.G., Bunger, A.P., Lecampion, B., Zhang, X., Chen, Z.R., et al., 2009. Measuring hydraulic fracture growth in naturally fractured rock. SPE 124919. SPE Annual Technical Conference and Exhibition, New Orleans, Louisiana, USA, 4–7 October.
Jorden, J.R., Shirley, O.J., 1966. Application of drilling performance data to overpressure detection. Journal of Petroleum Technology 18 (11), 8. Available from: https://doi.org/10.2118/1407-PA. Society of Petroleum Engineers.
Kresse, O., Weng, X., Chuprakov, D., Prioul, R., Cohen, C., 2013. Effect of flow rate and viscosity on complex fracture development in UFM model, In: Effective and Sustainable Hydraulic Fracturing.
Lade, P., 1977. Elasto-plasto stress-strain theory for cohesionless soil with curved yield surfaces. International Journal of Solids and Structures 13, 1019–1035.
Lai, J., Wang, G., Fan, Z., Wang, Z., Chen, J., Zhou, Z., et al., 2017. Fracture detection in oil-based drilling mud using a combination of borehole image and sonic logs. Marine and Petroleum Geology 84, 195–214.
Li, L., Tan, J., Tan, Y., Pan, X., Zhao, Z., 2022. Chapter Eight - Microseismic analysis to aid gas reservoir characterization. In: Wood, D.A., Cai, J. (Eds.), The Fundamentals and Sustainable Advances in Natural Gas Science and Eng, Sustainable Geoscience for Natural Gas Subsurface Systems, Vol. 2. Gulf Professional Publishing, pp. 219–242.
Maxwell, S.C., Rutledge, J., Jones, R., Fehler, M., 2010. Petroleum reservoir characterization using downhole microseismic monitoring. Geophysics 75, 129–137. 75A.
McGuire, W.J., Sikora, V.J., 1960. The effect of vertical fractures on well productivity. Transactions of the AIME 219, 401.
McNally, G.H.N., 1987. Estimation of coal measures rock strength using sonic and neutron logs. Geoexploration 24, 381–395.
Nolte, K.G., 1991. Fracturing pressure analysis for nonideal behavior. Journal of Petroleum Technology 43 (2), 210–218.
Nolte, K.G., Smith, M.B., 1981. Interpretation of fracturing pressures. Society of Petroleum Engineers Journal 33 (9), 1767–1775.
Nordgren, R.P., 1972. Propagation of a hydraulic fracture. SPE Journal 306–314. Trans. AIME, 253.
Perkins, T.K., Kern, L.R., 1961. Widths of hydraulic fractures. Journal of Petroleum Technology 13, 937–949.
Prats, M., 1961. Effect of vertical fractures on reservoir behavior-incompressible fluid case. SPE Journal 1 (2), SPE-1575-G.
Rehm, B., McClendon, R., 1971. Measurement of formation pressure from drilling data. Fall Meeting of the Society of Petroleum Engineers of AIME, 3–6 October, New Orleans, Louisiana, USA. Society of Petroleum Engineers. doi: 10.2118/3601-MS.
Rui-Zhao, Y., Zheng-Guang, Z., Wei-Jun, P., Yu-Bo, G., Zhan-Gang, W., Xi-Qin, Z., 2013. Integrated application of 3D seismic and microseismic data in the development of tight gas reservoirs*. Journal of Applied Geophysics 10 (2), 157–169.
Schlumberger Limited, 2015. Techlog 2015.3: User Guide. Schlumberger software.
Shehata, A., Aly, A., Ramsey, L., 2010. Overview of Tight Gas Field Development in the Middle East and North Africa Region. Society of Petroleum Engineers, SPE 126181.
Sneddon, I.N., 1946. The distribution of stress in the neighbourhood of a crack in an elastic solid. Proceedings of the Royal Society of London. Series A, Mathematical and Physical Sciences 187, 229–260.

Valko, P., Economides, M.J., 1995. Hydraulic Fracture Mechanics. John Wiley & Sons, p. 298.
Veatch Jr., R.W., 1983. Overview of current hydraulic fracturing design and treatment technology-part 1. Journal of Petroleum Technology 35 (4), 677−687. SPE-10039-PA.
Wang, H., 2019. Hydraulic fracture propagation in naturally fractured reservoirs: Complex fracture or fracture networks. Journal of Natural Gas Science and Engineering 68, 102911.
Weng, X., Kresse, O., Cohen, C., Wu, R., Gu, H., 2019. Modeling of hydraulic fracture network propagation in a naturally fractured formation. SPE 140253, paper presented at the SPE Hydraulic Fracturing Technology Conference, The Woodlands, Texas, USA, 24−26 January.
White, A.J., Traugott, M.O., Swarbrick, R.E., 2002. The use of leak-off tests as means of predicting minimum in situ stress. Petroleum Geoscience 8, 189−193.
Wiebols, G.A., Cook, N.G.W., 1968. An energy criterion for the strength of rock in polyaxial compression. International Journal of Rock Mechanics and Mining Sciences 5, 529−549.
Wu, K., Olson, J.E., 2014. Mechanics analysis of interaction between hydraulic and natural fractures in shale reservoirs. Unconventional Resources Technology Conference. Denver, Colorado, USA, 25−27 August 2014. RTeC: 1922946.
Xu W, Calvez, J.L., Thiercelin, M., 2009. Characterization of hydraulically-induced fracture network using treatment and microseismic data in a tight-gas formation: a geomechanical approach." SPE paper 125237 presented at the 2009 SPE Tight Gas Completions Conference, San Antonio, TX, USA, 15−17 June.
Yamamoto, K., 2003. Implementation of the extended leak-off test in deep wells in Japan. In Sugawara, K. et al. (Eds.), Proceedings of the Third International Symposium on Rock Stress Rs, Kumamoto "03, Rotterdam (Balkema), 225−229.
Zhang, J., 2011. Pore pressure prediction from well logs: methods, modifications, and new approaches. Earth-Science Reviews 180 (1), 50−63.
Zhang, J., Roegiers, J.-C., 2010. Discussion on "Integrating borehole-breakout dimensions, strength criteria, and leak-off test results, to constrain the state of stress across the Chelungpu Fault, Taiwan.". Tectonophysics 492, 295−298.
Zhang, J., Yin, S., 2017. Real-Time Pore Pressure Detection: Indicators and Improved Methods. Geofluids.
Zhang, J.J., Zhang, Y., Yin, S., 2018. PKN solution revisit: 3-D hydraulic fracture size and stress anisotropy effects. Rock Mechanics and Rock Engineering 51 (2), 653−660.
Zhang, S., Yin, S., Wang, F., Zhao, H., 2017. Characterization of in situ stress state and joint properties from extended leak-off tests in fractured reservoirs. International Journal of Geomechanics 17 (3), 1−12.
Zoback, M.D., 2007. Reservoir Geomechanics, first ed Cambridge University Press, New York, p. 464.

CHAPTER 11

Geology knowledge in the engineering applications of tight gas sandstones

11.1 Introduction

Hydrocarbon is produced out of rocks and therefore understanding the geology of reservoir rocks and in general porous media is necessary for engineering the application of tight gas sandstones. Geology knowledge is fundamental in different phases of tight sandstone reservoirs exploration and production including resources evaluation, drilling, stimulation, completion, and lifetime surveillance. An accurate geological model integrating geological, petrophysical, and seismic data aids in reserve evaluation, geomechanical modeling, and designing a satisfactory hydraulic fracturing treatment. Many drilling risks and costs can be minimized through sound geological knowledge. A comprehensive mechanical earth model (MEM) of tight sands is critical to overcoming wellbore instability issues, casing design, and designing a hydraulic fracturing job. Geology knowledge is also beneficial in lifetime surveillance of tight sands such as production logging after a declining flow rate.

11.2 Geological considerations in exploratory and drilling phase of tight gas sandstones

Various sources of geosciences data resulting from stratigraphic analysis, facies, depositional setting, and diagenesis studies, well logs interpretation, analysis of core data, and rock elastic properties combined with 2D/3D seismic data (e.g., acoustic impedance, seismic attributes) are integrated into a geological model for tight gas sandstones reservoir characterization. Having accurate geological knowledge of rock properties reduces the drilling risks and costs during the exploration and development phase of tight gas sandstones.

11.2.1 Well azimuth

The drilling well azimuth controls the wellbore stability and casing collapse issues in a tight gas field. Stress regimes can be extensional

Reservoir Characterization of Tight Gas Sandstones
DOI: https://doi.org/10.1016/B978-0-323-90180-2.00004-9

compressional and shear. In complex stress regimes, a combination of the extensional or compressional stresses with the shear stresses exists. In the extensional stress regime that is associated with the normal faulting system, vertical stress is the dominant stress (vertical stress > maximum horizontal stress > minimum horizontal stress). In the compressional stress regime that is associated with the reverse faulting system, maximum horizontal stress is the dominant stress (maximum horizontal stress > minimum horizontal stress > vertical stress). The shear stress regime is associated with a strike-slip faulting system (maximum horizontal stress > vertical stress > minimum horizontal stress).

Generally, vertical wells are more stable than deviated or horizontal wells. For the case of directional drilling, a good azimuth and stress regime has a major control on borehole stability and minimizes drilling risks and costs.

In the extensional stress regime, drilling parallel to the minimum horizontal stress is the safest direction, while drilling parallel to the maximum horizontal stress is the worst direction that is associated with more washouts and breakouts (Fig. 11–1A).

In the compressional stress regime, drilling parallel to the maximum horizontal stress is the safest direction, while drilling parallel to the minimum horizontal stress is the riskiest direction that is associated with more washouts and breakouts (Fig. 11–1B).

11.2.2 Overpressure zones

Before drilling a tight gas sandstone, countermeasures must be taken to deal with the overpressure and underpressure zones of a tight gas sandstone reservoir. Different stages of overpressure and underpressure happen in a tight gas sandstone. Once a mature source rock charges an encased low permeability sandstone causes an overpressure system. Leaking gas from the up-dip margins of an overpressured zone causes an underpressure margin around the overpressure zone (Burnie et al., 2008; Meckel and Thomasson, 2008; Zee et al., 2016).

11.2.3 Natural fractures

The presence of natural fractures plays an important role in the gas production enhancement but they can also cause an increase in the water production as well. Moreover, natural fractures are the main cause of mud loss and lost circulation. Interpreting 2D and 3D seismic based on seismic

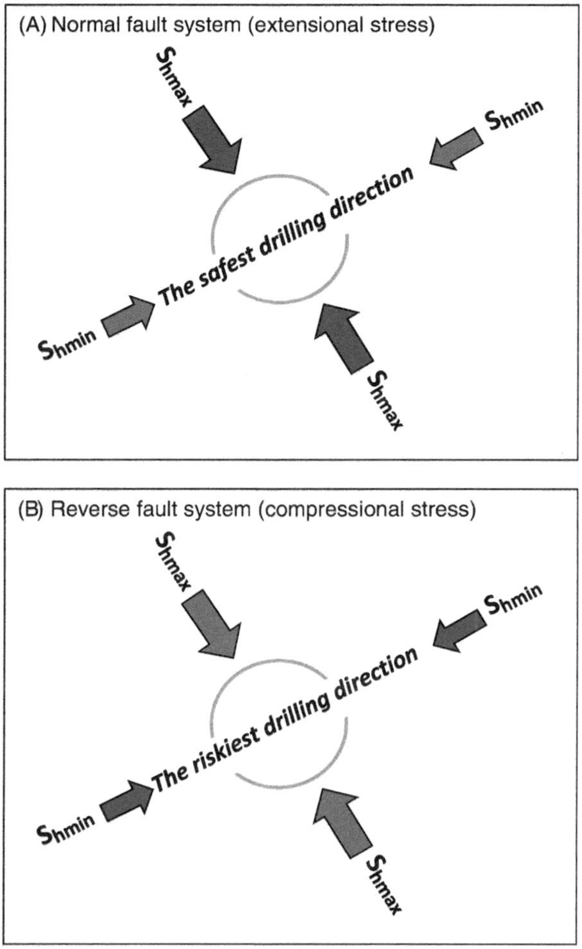

Figure 11–1 The safest and riskiest drilling directions in different stress regimes. In the extensional stress regime drilling parallel to the minimum horizontal stress is the safest direction from a borehole stability point of view, while drilling parallel to the maximum horizontal stress is the riskiest azimuth. In the compressional stress regime, drilling along the maximum horizontal stress is the safest direction and vice versa. *Originally generated for this chapter.*

attributes such as coherency, curvature and ant tracking, a distribution model of faults and fractures can be generated. Studying the faults and fractures based on the seismic data, regional studies and drilled formations is an important consideration in the well planning and geosteering of explorational or development wells in tight gas sandstones.

11.2.4 Lithological complexities

Formation damage, borehole washout, and abrasion of the drilling bit are the main drilling challenges in tight gas sandstones. Due to the geological complexities such as the tectonic activities, principle stresses orientation, and the presence of swelling clay minerals, they are subjected to borehole failures such as washouts and breakouts. The main lithological constituents of tight gas sandstones are sandstone, siltstone, shale, and mudstone. Some clay minerals such as montmorillonite and smectite are capable of swelling and washout when they are subjected to water-based mud filtrate. Two types of swellings occur including crystalline and osmotic. Osmotic swelling is caused when the smectite group comes into contact with drilling and completion fluids. In such cases, the water-based drilling fluid should not be used to minimize borehole collapse. Therefore the most important consideration of drilling a well in a tight gas sandstone is drilling a gage hole. This will ensure obtaining a complete suite of open-hole well logs. In tight reservoirs, it is difficult to have an accurate estimation of clay volume, effective porosity, and water saturation. This issue will become more problematic in case of drilling a washout or out of gage well making it difficult to discriminate between pay zones from nonpay intervals. Besides well logging issues in washout intervals, a primary cementing job will not be adequately obtained causing problems in zonal isolation and having cement squeezed wells prior to running formation testers or pumping fracturing fluid.

11.2.5 Underbalanced drilling

Another concern of tight gas sandstones is underbalanced drilling to maximize the penetration rate of the drill bit or to reduce the invasion of mud filtrate. In case of having washouts or breakouts, underbalanced drilling will incur a lot of damages and costs due to inaccurate well logging or inadequate primary cementing job. Therefore the best strategy is to have near-balanced drilling to reduce the washouts and uncertain well log readings. The alternative methods of drilling in tight gas sandstones are conventional drilling, casing while drilling (CwD), coiled tubing drilling (CTD), overbalanced drilling (OBD) and managed pressure drilling (MPD). The followings are the relative advantages and disadvantages of the different drilling scenarios in tight gas sandstone reservoirs (Pilisi et al., 2010; Zee et al., 2016).

Conventional drilling: stuck pipe may increase, lower formation damage, lower rate of penetration.

Casing while drilling: lower stuck pipe, little formation damage, little rate of penetration.

Coiled tubing drilling: no effect on drilling problems, lower formation damage, higher rate of penetration due to smaller borehole size.

Overbalanced drilling: higher stuck pipe, lower formation damage, lower rate of penetration, kick detection.

Underbalanced drilling: lower stuck pipe, higher formation damage, higher rate of penetration, kick detection.

Managed pressure drilling: Much lower stuck pipe, higher formation damage, higher rate of penetration, kick detection.

11.3 Geological considerations in the production and stimulation of tight gas sandstones

The completion of tight gas wells consists of well testing, cementation of production casing, hydraulic fracturing, and installation of tubing and downhole production equipment. Stage count, diversion technique (chemical or mechanical), perforation design, fluid, and proppant selection are the important considerations in the completion of tight sandstones (Holditch and Bogatchev, 2008; Zee et al., 2016). In this section, the main geological parameters that need to be considered in the production and stimulation of tight gas sandstones are outlined.

11.3.1 Reservoir geometry

From the geological point of view, two types of tight sandstone geometries occur in sedimentary basins including stacked tight sandstones and blanket tight sandstones (Fig. 11–2). Stacked tight sandstones are thick and form in braided rivers system, deltas, or turbidites, while blanket tight sands are usually extensive and can either be thin or thick. In order to enhance the performance of stimulation, vertical wells need to be completed in stacked tight sandstones and horizontal wells in thin blanket sandstones (Wei & Xu, 2015; Zee et al., 2016). However, horizontal wells in stacked sandstones may show successful production rates. Therefore an accurate understanding of depositional setting and rock properties along with a realistic geological model representing the geometry and size of sand bodies are prerequisite inputs for the selection of drilling scenarios and the development of tight gas sandstones.

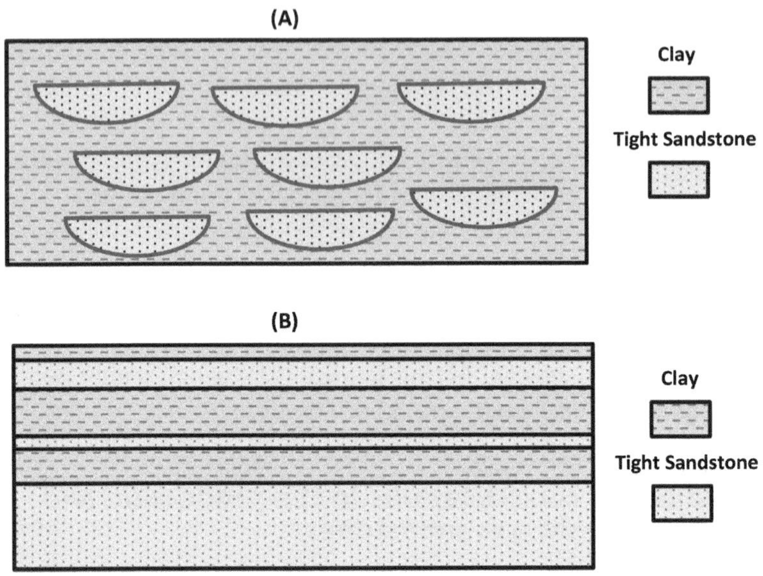

Figure 11–2 Stacked tight sandstones (A) and blanket tight sandstones (B). *Originally generated for this chapter.*

Although resources evaluation of tight gas and gas shale reservoirs are different, their drilling, design, and completion are similar (Kennedy et al., 2012). Many challenges of the tight gas sandstones are relevant to the poor connectivity between sand bodies due to low matrix permeability and the small size of reservoir geometry. The completion of a tight gas reservoir would be easier than other unconventional plays such as gas shales as they contain a much higher content of quartz and hence show a more brittle behavior causing a more success of hydraulic fracturing job. The horizontal and vertical wells drilled in a tight gas sandstone need to be stimulated successfully to have commercial gas flow rate and production. Most of the time, a single hydraulic fracturing treatment can address the production issues of multilayered stacked tight gas sandstones. As shown in Fig. 11–3, if a thick barrier (e.g., a shale layer) separates two productive zones and causes pressure contrast between them, multiple fracture treatments need to be carried out (Holditch, 2006, Zee et al., 2016). The geometry of a hydraulic fracture is directly controlled by the simultaneous interaction of elastic, geological, and reservoir properties of tight sandstones (Gupta and Duarte, 2016; Verma & Ojha, 2021). Depending on the tight formations' mineralogy, elastic and petrophysical properties, choosing the optimum fracturing treatment and pumping rates

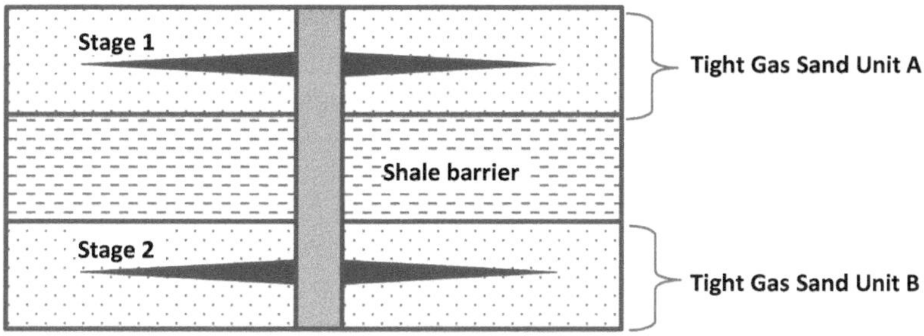

Figure 11–3 A schematic model of two-stage hydraulic fracturing treatment for two tight gas sandstone units separated by a thick shale barrier. *Originally generated for this chapter.*

controls the length and width of the hydraulic fractures. For example, in some tight formations, slickwater creates larger fractures, while linear gel treatment creates wider fractures and vice versa.

11.3.2 Well spacing

Compared to the conventional high permeability reservoirs, tight gas reservoirs development requires more wells, more data, and more manpower. The ultimate gas recovery of tight gas sandstones is dependent not only on the geological and petrophysical characteristics of rocks but also on the well spacing and well drainage area. The well completion and stimulation scenarios of tight gas sandstones are controlled by other factors such as the number of wells, their spacing, gas-bearing interval thickness, and drainage area. A well completed in a tight gas sandstone will produce less gas than a conventional reservoir over a long period of time. Therefore in order to have commercial production from tight gas sandstones, more wells with close spacing are required to be completed and stimulated. On the other hand, the drainage area is closely related to the well spacing and therefore an optimized well spacing aids in a more successful well stimulation and completion jobs.

11.3.3 Well azimuth

In the extensional stress regime that is associated with the normal faulting system, on-azimuth wells are parallel to the minimum horizontal stress. Off-azimuth wells are perpendicular or deviated from the minimum horizontal stress direction.

On-azimuth wells show satisfactory hydraulic fracturing and production rates compared to the off-azimuth wells. This shows that hydraulic fracturing performance and the resulting flow rate change as a function of drilling azimuth. Stephenson et al. (2018) with a study from two wells drilled in the Duvernay shale oil play of Canada concluded that production in the on-azimuth well is doubled compared to the 45 degrees off-azimuth well.

11.3.4 Pre-existing natural fractures

Hydraulic fracturing in a tight gas sandstone with pre-existing natural fractures can effectively enhance the fracture network and improve the gas flow rate. This way, hydraulic fractures could stimulate the adjacent natural fractures network and increase the width of the stimulated fractures (Kresse et al., 2013). In this regard, the results of formation image logs interpretation (e.g., FMI, UBI) are essential for being familiar with target formation fractures prior to designing the fracturing job. A base map or model of fracture intensities serves as the basis for discrete fracture network modeling. A hydraulic fracture may cross or open a natural fracture. When a hydraulic fracture intersects with a natural fracture, three cases of direct crossing, offset crossing, and arresting can happen (Fig. 11–4). As is seen, for the case of vertical wells drilled in reservoirs with pre-existing natural fractures, a larger difference between the SH_{max} and Sh_{min} would result in crossing the hydraulic fracture with the natural fracture plane and creating a planar fracture (Chuprakov et al., 2010; Kresse et al., 2013). In this case, the natural fracture is always kept close. In the circumstance of a lower difference between the minimum and maximum horizontal stresses, the hydraulic fracture follows the path of natural fractures. In this state,

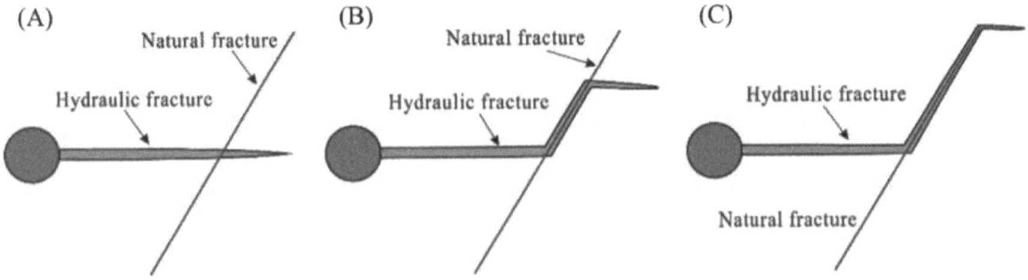

Figure 11–4 Possible cases of intersecting a hydraulic fracture with a natural fracture. (A) Direct crossing: the HF crosses the NF and propagates in the same direction. In this scenario, natural fracture stays close; (B) Offset crossing: the HF deflects into the natural fracture and kicks off at some weak points before reaching the NF tip; (C) Arresting or no crossing: the HF is arrested and propagates off the terminal of the natural fracture. In scenarios (B) and (C), the hydraulic fracture opens the natural fracture (Zhang et al., 2018).

the hydraulic fracture could completely be arrested or deviated through offset paths (Dahi-Taleghani and Olson, 2013; Wu & Olson, 2014).

11.3.5 Petrophysical properties

As with the conventional reservoirs, the main petrophysical properties of tight gas sandstones are porosity, permeability, the volume of shale, and water saturation. However, due to the critical importance of small permeability and porosity values, care must be taken in their accurate calculation, otherwise, pay zones can be missed or wrongly interpreted as nonpay zone. Apart from the permeability, other petrophysical parameters can accurately be calculated from good quality well logging data. Although there are some empirical correlations to estimate sandstones permeability, their results are highly uncertain as such formulas and their associated parameters are established for specific formations that cannot work everywhere. In this regard, using unconventional good logs such as nuclear magnetic resonance log (NMR) and its calibration with the core-derived permeability reduces uncertainty in permeability estimation. The candidate interval for hydraulic fracturing should have a moderate to high porosity and permeability. Ultra-low permeability and very low porosity zones cannot be successfully stimulated for cost-effective production.

11.3.6 Elastic rock properties

Determining the practicability of a target tight sandstone is highly dependent on the rock elastic moduli such as Poisson's ratio, Young's modulus, UCS, and brittleness index (BI). Fractures easily propagate in brittle formations, while in less brittle formations it is hard to overcome minimum horizontal stress and create efficient fractures. In order to propagate long, wide, and efficient fractures, the candidate interval chosen for hydraulic fracturing operation should have a higher brittleness index (BI), a higher *UCS*, a higher Young's modulus, and a lower Poison's ratio. Of course, petrophysical parameters should also be considered in candidate intervals selection. For example, an ultra-tight interval with a very high USC, a very high Young's modulus, and a very low Poisson's ratio is not a good candidate. Instead, a brittle tight interval but with enhanced porosity and permeability is preferred to ultra-tight intervals. Three popular approaches for BI calculation are mineral-based, elastic-based, and strength-based.

In the mineral-based approach, it is assumed that the presence of ductile minerals such as clays, micas, and feldspars reduces the BI of the tight sandstones. Therefore increasing the ductility of a tight formation will make it difficult to propagate hydraulic fractures. Quartz is a brittle mineral, while clays, feldspar, mica, and rock fragments are ductile. The mineralogical BI is expressed as follows.

$$BI = \frac{Q}{C + F + M + RF} \times 100 \tag{11.1}$$

where Q is quartz, C is clay, F is feldspar and RF is rock fragment in the fraction.

It should be noted that rock fragments are divided into sedimentary (SRF), volcanic (VRF), and metamorphic (MRF). Some rock fragments such as chert and limestone are brittle and hence must be excluded from the RF content and be added to the numerator of Eq. (11.1).

The elastic-based BI is calculated from the Poisson's ratio and Young's modulus as follows (Mathia et al., 2016).

$$BI = \frac{1}{2} \times (E + \upsilon_{brit}) \tag{11.2}$$

in which

$$E_{brit} = \frac{E - E_{min}}{E_{max} - E_{min}} \tag{11.3}$$

$$\upsilon_{brit} = \frac{\upsilon - \upsilon_{min}}{\upsilon_{max} - \upsilon_{min}} \tag{11.4}$$

where E_{brit} and υ_{brit} are the normalized Young's modulus and normalized Poisson's ratio, $E_{max}$and E_{min} are the maximum and minimum Young's modulus, $\upsilon_{max}$and υ_{min} are the maximum and minimum Poisson's ratio, E and υ are Young's modulus and Poisson's ratio for a given depth in an interval of interest.

The strength-based BI is calculated from the uniaxial compressive strength

(σ_C) and Brazilian tensile strength (σ_T) as follows (Altindag, 2003).

$$BI = \frac{\sigma_C - \sigma_T}{2} \tag{11.5}$$

$$BI = \frac{\sigma_C - \sigma_T}{\sigma_C + \sigma_T} \tag{11.6}$$

Eqs. (11.5) and (11.6) can be normalized to fall in the range of zero and 1. Roughly, the tensile strength is one-tenth of the compressional strength ($\sigma_T = 0.1 \times \sigma_C$). However, it needs to be calculated accurately based on laboratory test data to have a realistic estimation of the strength-based BI.

Stresses heterogeneity: The differences between maximum and minimum horizontal stresses affect the propagation of the hydraulic fracture. A larger difference between σ_H and σ_h generates longer and more planar hydraulic fractures. However, tight formations with a low difference between S_H and S_h create shorter and more complicated hydraulic fractures (Daniels et al., 2007; Dohmen et al., 2017; Zhang, 2019). The difference between the horizontal stresses can be quantified by the following equations proposed by Cruikshank et al. (1991) and Renshaw and Pollard (1994), respectively.

$$R = \frac{\sigma_H - \sigma_h}{p_t + \sigma_h} \tag{11.7}$$

$$R = \frac{\sigma_H - \sigma_h}{kT_0} \tag{11.8}$$

where R is stressed difference ratio, σ_H is maximum horizontal stress, σ_h is minimum horizontal stress, p_t is treatment pressure (pressure inside the fracture), T_0 is formation tensile strength, and k by default is taken as $\sqrt{2}$.

As shown by Renshaw and Pollard (1994), a lower horizontal stresses difference or R means higher tensile stress which in turn causes the tip of created hydraulic fracture curve to the adjacent natural fractures making a complex fracture network. Otherwise, a lower R causes straight and planar fracture (Fig. 11–5).

A tight gas sandstone with minimum horizontal stress less than its upper and lower barrier formations shall be a good candidate for hydraulic fracturing. Since a lower minimum horizontal stress means a lower formation breakdown pressure. In this case, the hydraulic fracture will only develop in the tight sandstone interval sandwiched between the upper and lower barriers. In addition to the stress barriers, petrophysical and elastic rock heterogeneities can terminate fractures growth. A high difference in the elastic properties (e.g., Young's modulus) between the upper and lower formation may contain (stop) hydraulic fractures. For example, a formation with high tensile strength overlying and underlying the tight sandstone can stop the fracture propagation. High tensile strength means high fracture initiation and formation breakdown pressures causing fracture propagation to be difficult. In this case, hydraulic fracture development follows the physics of fracture and mechanical stratigraphy.

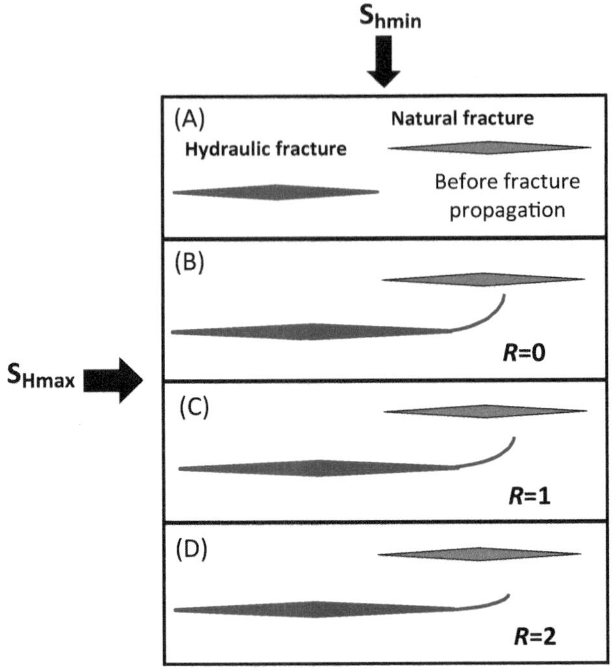

Figure 11–5 The effect of horizontal stresses difference (*R*) on fracture propagation path. A hydraulic fracture is created adjacent to a natural fracture (A). Increasing the horizontal stresses difference cause the created hydraulic fracture chooses a curved path toward adjacent natural fractures making a complex fracture network. *Modified after Renshaw, C.E., Pollard, D.D., 1994. Are large differential stresses required for straight fracture propagation paths? Journal of Structural Geology 16, 817–822.*

In addition to the stresses heterogeneity and rock elastic properties, the pumping rate and fracturing fluid rheology can also affect the propagation pattern of hydraulic fractures in tight sandstones with pre-existing natural fractures. It is worth mentioning that some failures in hydraulic fracturing jobs are not only related to the uncertainties in the mechanical earth model but they can also be caused by the inappropriate design of fracturing fluid.

Due to the low permeability nature of these reservoirs, performing pressure transient testing is a challenging issue making it difficult to acquire reservoir dynamic properties. Using the detailed and advanced tools it is feasible to measure formation pressure in a microdarcy rock accurately to determine drainage radius and optimal well spacing. The drilling, casing, and perforation can perfectly be carried out, but they will not produce successfully until an optimal treatment is pumped into the well. Gas flow rate, water production, artificial lift, tubing design, and casing diameter are the important considerations in the production stage of tight gas sandstones (Zee et al., 2016).

11.3.7 Stresses change due to depletion

The production and depletion of tight sandstones reservoir reduce the minimum and maximum horizontals stresses. A decrease in the minimum horizontal stress means a lower formation breakdown pressure and fracture propagation pressure is required. Accordingly, hydraulic fractures in a depleted producing well will be more easily generated and propagated. Similarly, re-fracturing a depleted well can improve the creation of two-wing symmetrical fractures (Dohmen et al., 2017; Zhang, 2019).

11.4 Geological considerations in lifetime surveillance of tight gas sandstones

After hydraulic fracturing treatment of tight gas sandstones, it is necessary to monitor the performance of the stimulation, fluid composition, and long-term pressure build-up (PBU). Apart from pumping rates, propping agent, and fracture treatment viscosity, fracture growth is highly dependent on the geological, petrophysical, and elastic properties of the tight gas formation. Fractures are easily developed in brittle formations and can form fracture networks that enhance the production rates effectively, while they cannot penetrate easily in ductile formations to create a network. Different types of fracture growth created after hydraulic fracturing treatment including planar, planar-coupled, complex planar, and network are shown in Fig. 11–6.

For this purpose, permanent downhole gauges, also called PDHG, are used to provide the above-mentioned information. Tight gas sandstones production is commingled from multiple zones and in many cases, most of the production is contributed to a few zones. In the same stratigraphic unit, similar high production zones may indicate spatially correlated sand layers. Integration of gamma-ray log and production data illustrates the lateral connectivity of sand bodies.

In order to understand the wellbore inflow profile, production logging is performed after treatment and later in its lifetime. Accordingly, a stable wellbore environment is essential to acquiring a high-quality production log. The high initial flow rate of tight sandstones declines rapidly after production. Too early, this will in turn affect the tool string causing a poor quality production log. On

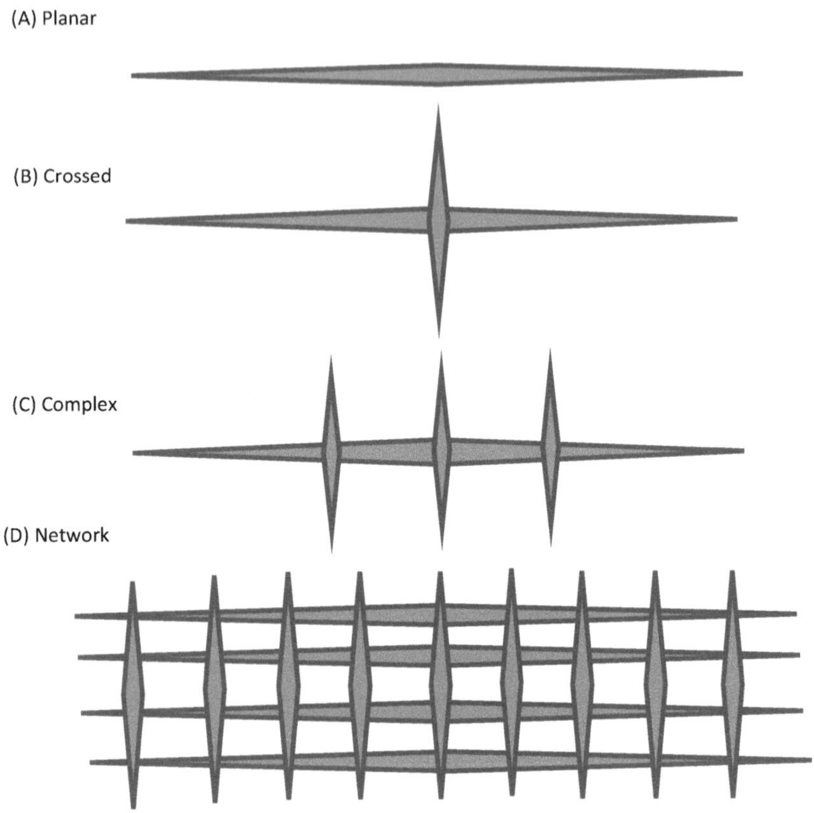

Figure 11–6 Different types of fracture growth after hydraulic fracturing including planar (A), crossed (B), complex (C), and network (D).

the other hand, too early logging does not provide an acceptable production log due to the production of fracturing fluids. Well shut-ins, plant shutdowns, and line pigging are the other causes of wellbore instability.

As a post-treatment analysis, the hydraulic fracture model needs to be tied to the dynamic simulation model for history matching and actual good production analysis. The rate transient analysis, also called production analysis, is coupled with the fracture modeling results to extract reservoir properties such as permeability, hydrocarbons-in-place, fracture conductivity, and fracture length.

References

Altindag, R., 2003. Correlation of specific energy with rock brittleness concepts on rock cutting. Journal of the Southern African Institute of Mining and Metallurgy (JSAIMM) 103 (3), 163–172.

Burnie, S.W., Maini, B., Palmer, B.R., Rakhit, K., 2008. Experimental and empirical observations supporting a capillary model involving gas generation, migration, and seal leakage for the origin and occurrence of regional gasifers. In: Cumella, S.P., Shanley, K.W., Camp, W.K. (Eds.), Understanding, Exploring, and Developing Tight Gas Sands, 2005 Vail Hedberg Conference, Vol. 3. AAPG Hedberg Series, pp. 29–48.

Chuprakov, D.S., Akulich, A.V., Siebrits, E., Thiercelin, M., 2010. Hydraulic fracture propagation in a naturally fractured reservoir. SPE 128715. In: Proceedings of the SPE Oil and Gas India Conference and Exhibition held in Mumbai, India, 20–22, January.

Cipolla, C.L., Warpinski, N.R., Mayerhofer, M.J., Lolon, E.P., Vincent, M.C., 2010. The relationship between fracture complexity, reservoir properties, and fracture-treatment design. SPE Production & Operations 25 (04), 438–452. SPE-115769-PA.

Cruikshank, K., Zhao, G., Johnson, A., 1991. Analysis of Minor Fractures Associated With.

Dahi-Taleghani, Arash, Olson, Jon E., 2013. How natural fractures could affect hydraulic-fracture geometrySPE 167608 SPE Journal.

Daniels, J., Waters, G., LeCalvez, J., Lassek, J., Bentley, D., 2007. Contacting more of the Barnett Shale through and integration of real-time microseismic monitoring, petrophysics, and hydraulic fracture design. In: Proceedings of the SPE Annual Technical Conference and Exhibition, Anaheim, California, SPE 110562.

Dohmen, T., Zhang, J., Barker, L., Blangy, J.P., 2017. Microseismic magnitudes and b-values for delineating hydraulic fracturing and depletion. Society of Petroleum Engineers Journal 22 (5), 1624–1634.

Gupta, P., Duarte, C.A., 2016. Coupled formulation and algorithms for the simulation of non-planar three-dimensional hydraulic fractures using the generalized finite element method. International Journal for Numerical and Analytical Methods in Geomechanics 40 (2016), 1402–1437.

Holditch, S.A., 2006. Tight gas sands. Journal of Petroleum Technology (June), 86–93.

Holditch, S.A., Bogatchev, K.Y., 2008. Developing tight gas sand adviser for completion and stimulation in tight gas sand reservoirs worldwide. paper spe 114195, presented at the CIPC/SPE gas technology symposium 2008 joint conference, Calgary Alberta, Canada, 16–19, June. joints and faulted joints. Journal of Structural Geology 13 (8), 865–886.

Kennedy, R.L., Knecht, W.N., Georgi, D.T., 2012. Comparison and contrasts of shale gas and tight gas developments, North American experience and trends. In: Proceedings of the SPE Saudi Arabia Section Technical and Exhibition, Al-khobar, Saudi Arabia, 8–11 April, SPE Paper 160855.

Kresse, O., Weng, X., Chuprakov, D., Prioul, R., Cohen, C., 2013. Effect of flow rate and viscosity on complex fracture development in UFM model. In: Effective and Sustainable Hydraulic Fracturing.

Mathia, E., Ratcliffe, K., Wright, M., 2016. Brittleness index- a parameter to embrace or avoid? In: Proceedings of the Unconventional Resources Technology Conference San Antonio Texas USA 1–3, August 2016, p. 10, URTeC: 2448745.

Meckel, L., M. Thomasson, 2008, Pervasive Tight-gas Sandstone Reservoirs: An Overview: AAPG Hedberg Series, No. 3, p. 13–27.

Pilisi, N., Wei, Y., Holditch, S.A., 2010. Selecting drilling technologies and methods for tight gas sand reservoirs. Society of Petroleum Engineers. Paper SPE 128191. In: Proceedings of the IADC/SPE Drilling Conference in New Orleans, Louisiana, USA, 2–4 February. <https://doi.org/10.2118/128191-MS>.

Renshaw, C.E., Pollard, D.D., 1994. Are large differential stresses required for straight fracture propagation paths? Journal of Structural Geology 16, 817–822.

Stephenson, B., Galan, E., Williams, W., MacDonald, J., Azad, A., Carduner, R., et al., 2018. Geometry and failure mechanisms from microseismic in the Duvernay Shale to explain changes in well performance with drilling azimuth. Society of Petroleum Engineers paper 189863.

Verma, M., Ojha, K., 2021. Application of response surface methodology for the optimization of viscosity of foam fracturing fluids for the unconventional reservoir. Journal of Natural Gas Science and Engineering 94, 104086.

Wei, Y., Xu, J., 2015. Development of liquid-rich tight gas sand plays - granite wash example. In: Ma, Y.Z., Holditch, S., Royer, J.J. (Eds.), Unconventional Resource Handbook: Evaluation and Development. Elsevier.

Wu, K., Olson, J.E., 2014. Mechanics analysis of interaction between hydraulic and natural fractures in shale reservoirs. In: Proceedings of the Unconventional Resources Technology Conference. Denver, Colorado, USA, 25–27 August 2014. RTeC: 1922946.

Zee, Ma, Y., Moore, W.R., Gomez, E., Clark, W.J., Zhang, Y., 2016. Chapter 14 - Tight gas sandstone reservoirs, part 1: overview and lithofacies. In: Zee Ma, Y., Holditch, S.A. (Eds.), Unconventional Oil and Gas Resources Handbook. Gulf Professional Publishing, pp. 405–427.

Zhang, J.J., 2019. Chapter 11 - Geomechanics applications in hydraulic fracturing. In: Zhang, J.J. (Ed.), Applied Petroleum Geomechanics. Gulf Professional Publishing, pp. 441–481.

Zhang, L., Zhou, J., Braun, A., Han, Z., 2018. Sensitivity analysis on the interaction between hydraulic and natural fractures based on an explicitly coupled hydrogeomechanical model in PFC2D. Journal of Petroleum Science and Engineering Engineering 167, 638–653.

Index

Note: Page numbers followed by "*f*" and "*t*" refer to figures and tables, respectively.

D

N

O

P

U

V

W

X

Y

Kolibri 360 GmbH
Ferdinand-Jühlke-Straße 7
99095 Erfurt, Deutschland
produktsicherheit@kolibri360.de